SpringerBriefs in Applied Sciences and Technology

SpringerBriefs in Thermal Engineering and Applied Science

Series Editor

Francis A. Kulacki, University of Minnesota, USA

For further volumes:
http://www.springer.com/series/10305

Sameer Khandekar • Krishnamurthy Muralidhar

Dropwise Condensation on Inclined Textured Surfaces

 Springer

Sameer Khandekar
Department of Mechanical Engineering
Indian Institute of Technology Kanpur
Kanpur, UP, India

Krishnamurthy Muralidhar
Department of Mechanical Engineering
Indian Institute of Technology Kanpur
Kanpur, UP, India

ISSN 2191-530X   ISSN 2191-5318 (electronic)
ISBN 978-1-4614-8446-2 ISBN 978-1-4614-8447-9 (eBook)
DOI 10.1007/978-1-4614-8447-9
Springer New York Heidelberg Dordrecht London

Library of Congress Control Number: 2013945394

Acknowledgements

We are thankful to our doctoral and master's students who worked with us in the areas of interfacial transport phenomena and phase-change. We have worked closely with Dr. Basant S. Sikarwar on most of the topics contained in this monograph. In addition, we would like to acknowledge the contributions of the following graduate students:

Nirmal Kumar Battoo
Gagan Bansal
Smita Agarwal
Sumeet Kumar

We have drawn liberally from their doctoral dissertation and master's thesis. The computer code developed by an earlier student, Dr. Trushar Gohil, was used for some parts of this research.

We thank Professor Animangshu Ghatak for introducing us to the subject of chemical texturing. Manoj Sharma, C. S. Goswami, and Sanjeev Chauhan helped us with condensation experiments reported in this work.

Financial support from funding agencies allowed us to make the apparatus and configure the measurement systems. We gratefully acknowledge the Department of Science and Technology, New Delhi, Board of Research in Nuclear Sciences, Mumbai, and the Ministry of Human Resource Development, New Delhi. Additional support from Center for Development of Technical Education (CDTE), IIT Kanpur, India, has been timely and is acknowledged.

We thank Professor Frank Kulacki, Series Editor, for providing us this opportunity and continued support.

We are grateful to our institute for the ambience it provides and our families for their endless patience.

Authors

Permissions

Figure	Source
2.1	Reprinted from Wu Y., Yang C., and Yuan X., (2001), Drop distribution and numerical simulation of dropwise condensation heat transfer, International Journal of Heat and Mass Transfer, vol. 44, pp. 4455–4464, with permission from Elsevier.
2.3	Reprinted from Bentley P. D., and Hands B. A., (1978), "The condensation of atmospheric gases on cold surfaces", Proc. of the Royal Society of London. Series A, Mathematical and Physical Sciences, vol. 359, pp.319–343, by permission of the Royal Society.
2.5	This figure was published in Song T., Lan Z., Ma X., and Bai T., (2009), Molecular clustering physical model of steam condensation and the experimental study on the initial droplet size distribution, International Journal of Thermal Sciences, vol. 48, pp. 2228–2236, Copyright © 2009 Elsevier Masson SAS. All rights reserved.
2.8	Reprinted Figure 2 with permission from Shi F., Shim Y., Amar J. G. (2005), Island–size distribution and capture numbers in three–dimensional nucleation: Comparison with mean–field behavior, Physical Review B, vol. 71, pp. 245411–245416, Copyright © 2005 from American Physical Society (http://link.aps.org/doi/10.1103/PhysRevB.71.2454 11).
2.18	Reprinted with permission from Leach R. N., Stevens F., Langford S. C., and Dickinson J. T., (2006), Dropwise condensation: experiments and simulations of nucleation and growth of water drops in a cooling System, Langmuir, vol. 22, pp. 8864–8872, Copyright©2006 from American Chemical Society.
2.20	Reprinted from Citakoglu E., and Rose J. W., (1969b), Dropwise condensation the effect of surface inclination, International Journal of Heat and Mass transfer, vol. 12, pp. 645–651, with permission from Elsevier.
2.21	Reprinted from Suzuki S., Nakajima A., Sakai M., Song J., Yoshida N., Kameshima Y., and Okada K., (2006), Sliding acceleration of water droplets on a surface coated with fluoroalkysline and octadecyltrimethoxysilane, Surface Science, vol. 600, pp. 2214–2219, with permission from Elsevier.
4.4	Reprinted from Leipertz A., and Fröba A. P., (2008), Improvement of condensation heat transfer by surface modifications, Heat Transfer Engineering, vol. 29 (4), pp. 343–356, with permission from Taylor & Francis Ltd. (http://www.tandf.co.uk/journals).

(continued)

(continued)

Figure	Source
4.6	Reprinted from Bhushan B., and Jung Y. C., (2011), Natural and biomimetic artificial surfaces for super-hydrophobicity, self-cleaning, low adhesion, and drag reduction, Progress in Materials Science, vol. 56, pp. 1–108, with permission from Elsevier.
4.7 and 4.8	Reprinted with permission from Cha T. G., Yi G. W., Moon M. W., Lee K. R., and Kim H. Y., (2010), Nanoscale patterning of micro-textured surfaces to control super-hydrophobic robustness, Langmuir, vol. 26(11), pp. 8319–8326, Copyright © 2010 from American Chemical Society.

Contents

Nomenclature

A	Surface area (m^2); suffixes sl for solid–liquid, lv for liquid vapor, cd for substrate average
C_p	Specific heat at constant pressure (W/kg K)
d_b	Base diameter of drop (m)
D	Diffusion coefficient of adatoms
$f_i(s/S)$	Scaling function for the island size distribution
f	Degree of roughness of substrate
F	Force (N); suffix σ for surface tension and g for gravity; Also, mass flux of monomers in the atomistic model
g	Acceleration due to gravity (m/s^2)
h_{lv}	Latent heat of vaporization (J/kg)
h	Heat transfer coefficient (W/m^2 K)
$\hat{i}, \hat{j}$ and $\hat{k}$	Unit vectors in x, y, and z directions
j_n	Average velocity of vapor molecules (m/s)
k	Thermal conductivity of condensate (W/m K)
l_{ij}	Distance between two nucleation sites i and j (m)
m	Mass of droplet (kg); also, mass of a vapor phase particle; Δm is the incremental mass condensed over a time step Δt
m''	Mass flux (kg/m^2 s)
M	Maximum size of unstable clusters
$\bar{M}$	Molecular weight of the condensing liquid (kg/kmol)
n_1	Number density of monomers (monomers/m^2)
n_d	Number of atoms/molecules in a drop of minimum radius

n_i Number density of critical clusters (i refers to that cluster size which does not decay but may change due to growth by the addition of clusters) (m^{-2})

n_j Number density of clusters containing j atoms/molecules (m^{-2})

n_s Number of clusters of size s at coverage ϑ

n_x Number density of stable clusters ($n_x = \sum_j n_j$ for all $j > i$) (m^{-2})

N Number of nucleation sites (cm^{-2}); N_f for a textured surface

N_A Avogadro number

p Pressure (N/m^2); v for vapor and l for liquid; sat for saturation

q'' Average heat flux (W/m^2)

q_d Surface heat transfer rate (W)

r Radius of drop (m); suffix b is for base radius

r_{min} Radius of thermodynamically smallest drop (m)

r_{cap} Capillary length, $\sqrt{\sigma/g(\rho_l - \rho_v)}$ (m)

r_{max} Size of the drop at instability due to fall-off (m)

r_{crit} Size of the drop at instability due to slide-off (m)

R Specific gas constant (J/kmol K); $\bar{R}$ is the universal gas constant

$S = \dfrac{\sum s \cdot n_s}{\sum n_s}$ Average island size; $s = 1, 2, 3, \ldots$

$t, \Delta t$ Time, time step (s)

T Temperature (K); l, v, and w are for liquid, vapor, and wall; sat for saturation

ΔT ($T_{sat} - T_w$)   Temperature difference between the saturated vapor and condensing wall (K)

u, v, w Velocity component in x, y, and z directions (m/s)

U Relative velocity between the wall and the drop (m/s); also terminal velocity

V Volume of the drop (m^3); c, i, and j are for the centroid and locations i and j

x, y, z Cartesian coordinates

X Characteristic distance for a graded surface

Z Fraction of surface covered by a stable cluster

Nondimensional Parameters

C_f Local skin friction coefficient ($2\tau_w/\rho U^2$)

$\bar{C}_f$ Average skin friction coefficient ($2\bar{\tau}_w/\rho U^2$)

Ja Jakob number, $\left(C_p/h_{lv}\right)_{ref}(T_{sat} - T_w)$

Nu	Nusselt number (hr_{cap}/k)
$(Nu)_{sd}$	Local Nusselt number ($h_{sd}d_b/k$)
$\overline{Nu}_{sd}$	Average Nusselt number ($\overline{h}_{sd}\,d_b/k$)
Pr	Prandtl number ($\mu C_p/k$)
Re	Reynolds number ($\rho U d_b/\mu$)

Dimensionless Quantities

$p/(\frac{1}{2}\rho U^2)$	Dimensionless pressure
$(T - T_w)/\Delta T$	Dimensionless temperature
u/U	Dimensionless velocity in x direction
v/U	Dimensionless velocity in y direction
w/U	Dimensionless velocity in z direction

Greek Symbols

α	Inclination angle of the substrate from horizontal (deg or rad)
α_l	Thermal diffusivity (m²/s)
β	Volumetric expansion coefficient (K⁻¹)
δ	Thickness of promoter layer (m)
δ_j	Decay constant of a cluster with j particles, s⁻¹
Γ	Progress velocity in the vapor phase
μ	Dynamic viscosity (Pa s)
ρ	Density (kg/m³); suffix l for liquid and v for vapor
σ	Surface tension (N/m); suffix lv, sv, and ls are the solid (s), liquid (l), and vapor (v) interfaces
$\hat{\sigma}$	Accommodation coefficient
σ_1	Capture rate of monomers by formation of dimers
σ_j	Capture number of clusters containing j atoms
$\tau_w, \overline{\tau}_w$	Local and average wall shear stress (N/m²)
θ	Contact angle (rad or °); subscript * refer to rough surfaces, w and c to Wenzel and Cassie states; adv and rcd to advancing and receding angles, respectively
ξ	Azimuthal angle (° or rad)

Abstract

Dropwise condensation is a heterogeneous phase-change process in which vapor condenses in the form of discrete liquid drops on or underneath a cold substrate. The heat transfer coefficient of dropwise condensation can be up to an order higher than film condensation and mixed-mode condensation, particularly with low-conductivity liquids. Therefore, it is of considerable interest in applications such as thermal power plants and condensing equipment. It is also of interest in the material enrichment of large molecular weight liquids. Dropwise condensation is complex process, involving drop formation at the atomic scale, growth of drops by direct condensation, coalescence of drops, drop instability and movement, followed by fresh nucleation. Hence, the dropwise condensation process is hierarchical in the sense that it occurs in a wide range of length and timescales. In addition, it depends on the thermophysical properties of the condensing fluid, physicochemical and thermal properties of the cold substrate, orientation of the cold substrate, surface texture, degree of subcooling, thermodynamic saturation conditions, and presence of noncondensable gases. As the driving temperature difference for the process is very small, experimental measurement of heat transfer coefficient in dropwise condensation is a challenging task. Against this background, a mathematical model of dropwise condensation process underneath an inclined surface is presented in this monograph. The model includes formation of drops at the atomic scale, growth by direct condensation, coalescence, gravitational instability including slide-off and fall-off, followed by fresh nucleation of liquid droplets. The stability criterion is developed as a force balance equation at the level of a drop. Transport parameters of a sliding drop are determined using a CFD model and presented in the form of correlations. Performing the simulation of the complete cycle of dropwise condensation, the spatiotemporal distribution of drops is obtained. Consequently, quantities such as instantaneous condensation pattern, area of coverage, wall friction, and heat transfer rates, as well as important time- and area-averaged wall heat fluxes are determined. The simulated condensation patterns are compared against experimentally recorded images. The model is also validated against wall heat fluxes reported in the literature. While applicable for a wide range of fluids such as water and liquid metals, the model is seen to be sensitive to surface texture, inclination, and saturation conditions.

Chapter 1
Introduction

Keywords Classification • Hydrophobic surface • Drop formation • Condensation cycle • Wettability • Contact angle • Hysteresis

1.1 Classification

Condensation involves change of phase from the vapor state to the liquid. It is associated with mass transfer, during which vapor migrates towards the liquid–vapor interface and is converted into liquid. Transport is driven by a pressure reduction that occurs at the phase boundary. Condensation process is initiated by a temperature difference, called subcooling, between the bulk vapor and the solid surface. Subsequently, energy in the form of the latent heat must be removed from the interfacial region either by conduction or convection. Apart from natural phenomena, condensation is an essential part of energy conversion, water harvesting, and thermal management systems. Improvement in heat and mass transfer during the phase-change process, therefore, can have beneficial effects.

Classification of the condensation process and the corresponding pictorial view of possible condensation patterns are depicted in Figs. 1.1 and 1.2, respectively. Homogeneous condensation occurs in free space in the absence of any foreign material. It takes place stochastically as result of fluctuations in the vapor molecules. Such a process is only occasionally seen and barely plays an important role in heat transfer devices. As an example, Fig. 1.2a shows satturated steam flowing in a pipe, where sudden expansion creates a favorable condition for the condensation of vapor. Heterogeneous condensation occurs when vapor condenses on or underneath the surface of any other material (either liquid or solid) or on spatially distributed nuclei, Fig. 1.2b–e. According to the type of condensing surface, heterogeneous condensation is divided into volume condensation and surface condensation. An example of volume condensation is formation of clouds, mist, or fog. Surface condensation takes place on or underneath a subcooled surface that is exposed to vapor. The resulting heat transfer coefficient is orders of

S. Khandekar and K. Muralidhar, *Dropwise Condensation on Inclined Textured Surfaces*, SpringerBriefs in Applied Sciences and Technology 11, DOI 10.1007/978-1-4614-8447-9_1, © Springer Science+Business Media New York 2014

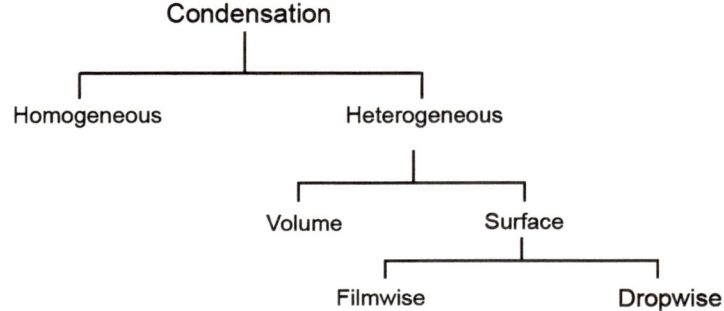

Fig. 1.1 Classification of condensation phenomena

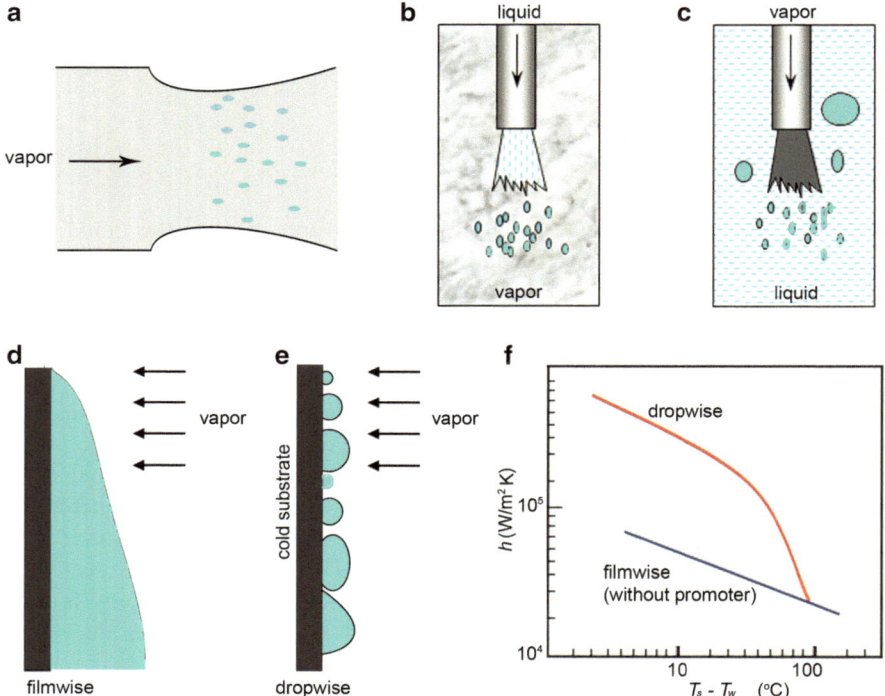

Fig. 1.2 Pictorial views of various types of condensation: (**a**) homogeneous condensation of steam due to pressure drop in a steam nozzle, (**b**) heterogeneous volume condensation, (**c**) heterogeneous condensation on liquid surface, (**d**) Filmwise condensation, and (**e**) dropwise condensation on a vertical cold substrate. (**f**) Comparison of dropwise and filmwise condensation heat transfer data for steam at atmospheric pressure

magnitude greater than the single-phase convective paradigm. Hence, it plays an important role in many heat transfer devices.

The phase-change process may result in either (1) the formation of a continuous film on the cold substrate (filmwise condensation) or (2) the formation of a droplet ensemble (dropwise condensation). There can be a mixed mode as well, having fuzzy overlapping characteristics of drops and a liquid film. The condensation form that is realized depends on the wettability of the surface, related to the free energy of the condensing wall and the surface tension of the condensate. Filmwise condensation occurs when the liquid wets the substrate while dropwise condensation takes place when the liquid does not have high affinity for the substrate. These processes are of interest from an engineering point of view as condensation occurs often in industrial equipment.

1.2 Filmwise Condensation

Filmwise condensation is preferred when the liquid wets the condenser surface, resulting in the complete coverage of the surface by a liquid film. It is commonly observed in various phase change heat transfer devices. The film is removed from the surface under the action of the gravity, acceleration, or other body forces and shear stresses due to vapor flow. The film renders a high thermal resistance to heat transfer and therefore, a relatively large temperature gradient prevails across it.

1.3 Dropwise Condensation

Vapor-to-liquid phase-change process in the form of discrete drops on or underneath a cold substrate is called dropwise condensation. It is realized when the condensate does not wet the substrate except at locations where well-wetted contaminant nuclei exist, Fig. 1.2e. The heat transfer coefficient during this process is an order of magnitude larger than for filmwise condensation, Fig. 1.2f. This makes dropwise condensation a very attractive mechanism for industrial applications.

Dropwise condensation begins with drop formation at preferred nucleation sites at the atomic scale. These droplets grow by direct condensation, up to a size of the order of the distance between neighboring nucleation sites. Beyond this point, coalescence among neighboring drops takes place and subsequent growth of drops occurs by the combination of direct condensation and coalescence. When a drop reaches a size, at which the body forces exceed surface tension holding it to the solid surface, the drop departs and sweeps the surface clear, permitting new nucleation sites to become available. Hence, coalescence and sliding droplets reexpose substrate area to provide a continuous source of nucleation sites. Overall, dropwise condensation is a quasi-cyclic process, as represented in Fig. 1.3. Several subprocesses of distinct length and timescales interact in space and time to form

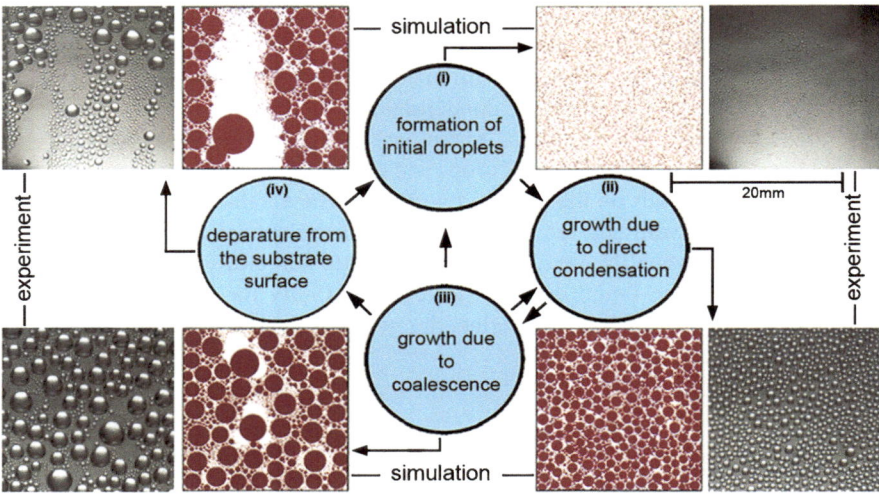

Fig. 1.3 Cycle of processes in dropwise condensation of vapor over an inclined cold substrate

a closed cycle of events. Various researchers (Mikic 1969; Griffith 1985; Tanasawa 1991; Rose 2002; Carey 2008) have confirmed that dropwise condensation is a complex phenomenon involving an interplay of several factors—from molecular level forces at the three-phase contact line of the droplets to the body forces acting on the condensing liquid droplets.

Schmidt et al. (1930) reported a high heat transfer coefficient for dropwise condensation. Ever since, it has been of considerable interest to many researchers because of it being a facile method of enhancing the efficiency of power generation units, thermal management systems, and water desalination systems. Special surfaces are required with strong dewetting characteristics (Carey 2008). In recent years, many research groups (Chen et al. 2009; Baojin et al. 2011; Ucar and Erbil 2012; Feng et al. 2012) have tried to create hydrophobic substrates by techniques such as advanced machining, nanotechnology, and thin films.

1.4 Understanding Dropwise Condensation

Industrial applications of dropwise condensation have not been very successful. This is because of the intricacies faced in controllability and long-term sustainability of the process on textured substrates. Its dependence on a large number of parameters such as nucleation site density, hydrophobicity, thickness of promoter layer, substrate orientation, degree of subcooling, and environmental conditions adds to the difficulty in modeling. In addition, many issues are unresolved. These include drop formation at the atomic scale, hierarchical phenomena, growth and coalescence mechanisms, contact angle hysteresis, dynamics of the three-phase contact line, instability of drops, and leaching.

Small changes in the surface morphology on a micro/nanoscale lead to changes in the droplet distribution, thereby affecting drop mobility. The overall temperature difference is small making measurement of heat transfer coefficient quite difficult. In addition, the statistical nature of droplet distribution in the ensemble contributes to the intricacy of analysis and interpretation. Reported heat transfer data of dropwise condensation invariably shows large scatter.

Dropwise condensation can be sustained only if the integrity of the condensing surface is maintained over long periods of time. In practice, condensing surface features are altered during growth, coalescence, and slide-off of the drops. Sustaining dropwise condensation for long time periods on engineered surfaces is a major challenge. Issues that require attention can be summarized as follows:

1. *Longtime sustainability*: The foremost is devising reliable means of promoting a cyclic dropwise condensation process. The ideal substrate has low thermal resistance, high durability, and low surface energy. With the advent of nano-technology, breakthroughs in thin film coating, physical and chemical texturing, and availability of superior experimental techniques, definite possibilities arise for sustaining dropwise condensation over a longer duration.
2. *Heat transfer measurement*: Experimental heat transfer investigations report widely scattered data because of inconsistency and difficulty in measurement of very low temperature differences.
3. *Substrate thermal conductivity*: The effect of thermal conductivity of the substrate on heat transfer during dropwise condensation is in controversy.
4. *Mechanism of dropwise condensation*: The interrelationship among the hierarchy of processes from the atomic scale to the drop is not addressed. A few fundamental questions remain unanswered. What part of the surface forms nucleation sites? What is the relation between the surface morphology and nucleation site density? How are drops distributed on or underneath the substrate? Is the equation determining minimum drop radius correct? How are critical sizes of drop at slide-off and fall-off calculated? Details of coalescence are unaddressed. Role of pressure and temperature fluctuations on leaching needs to be understood. Effects of substrate orientation, hydrophobicity, and surface energy gradient on heat transfer rates are of importance.
5. *Controllability*: Dropwise condensation depends on thermophysical properties of the condensing fluid, physicochemical properties of the cold substrate, its orientation, surface energy profile, subcooling, and saturation pressure. Thus, close control of dropwise condensation is difficult and one needs clarity in understanding the importance of parameters on various length and timescales.
6. *Multiscale phenomenon*: The overall mechanism of formation of a droplet on a textured surface involves varied length scales, from atomistic orders at early stage of nucleation, to scales affected by the body forces, while surface tension is important at intermediate scales. Thus, dropwise condensation can be understood only when a multiscale modeling approach is adopted.

1.5   Intermediate Steps in Dropwise Condensation

Dropwise condensation begins at an atomistic level in which vapor atoms impinge on the cold substrate. These individual atoms will form stable clusters which lead to microscopic droplets at a specific location on a surface. These grow by direct condensation of the vapor and by coalescence between droplets, until a certain size is reached. Drops then leave the surface by the action of body forces and vapor shear and reexpose the substrate area. Fresh nucleation occurs at the reexposed area and the complete condensation cycle begins at the atomic level once again. The atomistic model captures the initial stage of condensation, which leads to a stable cluster. Once a cluster is formed, bulk thermophysical properties of the liquid and physicochemical properties of substrate become relevant and start influencing growth. Dropwise condensation can be sustained if the condensate does not wet the cold surface. Figure 1.4 shows the schematic diagram of various types of solid–liquid drop interaction on a planar surface. Wetting characteristics can be established by the measurement of the apparent contact angle θ, specific to the choice of the liquid and the surface material (DeGennes 1985). It is defined as the angle between the tangents drawn at the liquid–vapor interface and the liquid–solid interface. The wettability of a surface by a liquid is a consequence of a combination of complex processes. Some of these originate at the microscale and can be understood in terms of surface chemistry and long-range van der Waals forces. Certain factors are purely statistical and may vary from sample to sample. These include wetting transitions and the pinning of the contact line. Fluid motion inside the droplet commences when it starts moving due to a force imbalance. The shape of the droplet will then depend on the principles of fluid dynamics as well. As a first step, the solid–liquid interaction in a drop may be characterized uniquely by the apparent contact angle θ and determined by measurements when the drop is in mechanical equilibrium. Liquid is said to wet a solid surface completely if it spreads over a considerable distance with a limiting value of $\theta = 0°$, Fig. 1.4a. If it retains a full spherical drop on contact with a solid surface, it is said to be fully nonwetting with contact angle $\theta = 180°$, Fig. 1.4c. In between the wetting and nonwetting regimes, there can be a situation when a liquid has a contact angle $0 < \theta < 180°$. This situation is known as partial wetting and the liquid has a finite

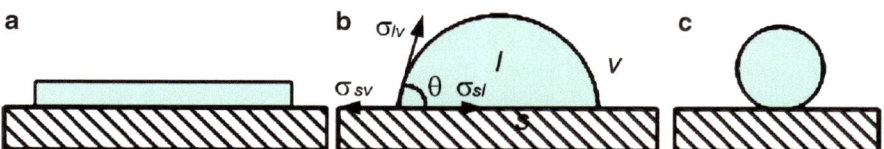

Fig. 1.4 Schematic representation of surface–drop interaction on a planar surface at the continuum scale. (**a**) Complete wetting ($\theta = 0$), (**b**) partial wetting ($0 < \theta < 180°$), and (**c**) nonwetting ($\theta = 180°$)

liquid–solid interface, as shown in Fig. 1.4b. In the real engineering context, most systems involning solids and liquids are invariably between the wetting and the nonwetting limits.

The contact angle contains details of the interactions at various interfaces including solid–liquid, liquid–gas, solid–gas and solid–liquid–gas. The adoption of the apparent contact angle simplifies analysis and helps understand the behavior of drops from a mechanics perspective. For a given liquid, a wide variety of substrates, natural and engineered, will produce a range of contact angles. These are classified as hydrophilic ($0° < \theta \leq 90°$), hydrophobic ($90° < \theta \leq 140°$) and superhydrophobic ($140° < \theta \leq 180°$) (Berthier 2008).

Several important phenomena in condensation rely on partial wetting of the solid substrate by the condensing liquid. Surface heterogeneities on the condensing substrate—chemical and topographical—have profound effect on the apparent contact angle and give rise to contact angle hysteresis and local pinning of the three-phase contact line. Based on the knowledge of contact angle, the behavior of a liquid drop on, or underneath a solid surface, is obtained. As the contact angle influences the equilibrium shape of the drop and hence its curvature, it can be related to interfacial tension and the surface energy distribution of the solid substrate.

1.5.1   Measurement of Contact Angle

Three-phase contact lines are formed when materials in different phases, e.g., solid, liquid, and gas (or vapor) intersect. Common examples are a liquid drop spreading on a solid surface or a liquid meniscus in a capillary tube. In the presence of the third phase (gas or vapor), a liquid spreading on a solid surface can reach two distinct equilibrium states. These are the following: (a) partial wetting and (b) complete wetting. The condition for static equilibrium of a triple contact line involving an ideal solid (perfectly smooth and chemically homogeneous), liquid, and a gas/vapor surrounding it is stated in the form of the classical Young's equation as

$$\sigma_{lv} \cos \theta = \sigma_{sv} - \sigma_{sl} \tag{1.1}$$

Here, the symbol σ_{ab} is the surface tension between phases a and b. Symbols s, l, and v in (1.1) stand for solid, liquid, and the gaseous phases respectively. The symbol θ is the apparent contact angle at each point of the solid–liquid boundary. Equation (1.1) holds for an ideally smooth solid surface with no chemical heterogeneities. Real solid surfaces depart from an ideal behavior since they are not perfectly smooth. In addition, their composition may also vary slightly with location. Molecules, atoms, or ions of other chemical species may be adsorbed on the surface. Effectively, the static contact angle turns out to be nonunique on real surfaces and can only be experimentally determined.

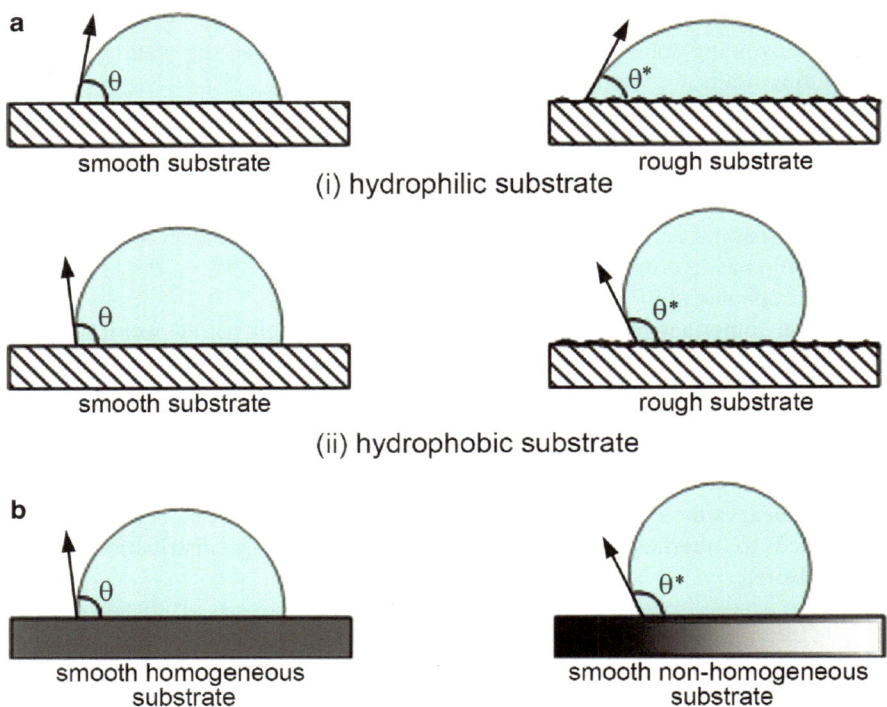

Fig. 1.5 Measurement of contact angle on a textured substrate. (**a**) Effect of roughness on contact angle (Wenzel Law) for a hydrophilic substrate and hydrophobic substrate. (**b**) Effect of chemical nonhomogeneity on local contact angle (Cassie-Baxter relation)

The experimentally observed contact angle depends on the way the surface is prepared, Fig. 1.5. One of the first attempts on understanding the influence of surface roughness on wetting is due to Wenzel (1936) who proposed the following relationship for the apparent contact angle

$$\cos \theta^* = f \cdot \cos \theta \qquad\qquad (1.2)$$

Here, θ^* is the apparent contact angle, f is the degree of roughness (with $f = 1$ for a smooth surface, $f > 1$ for a rough surface, and θ, the local apparent contact angle). Equation (1.2) embodies two types of behavior for rough surfaces. For hydrophilic behavior, we have $\theta^* < \theta$ since $f > 1$, as shwon in Fig. 1.5a-i. Likewise, for hydrophobic, we have $\theta^* > \theta$, as depicted in Fig. 1.5a-ii. Many researchers (Huh and Mason 1977; Leger and Joany 1977; DeGennes 1985) have shown that a wetting experiment is extremely sensitive to heterogeneities of the solid surface. Shibuichi et al. (1996) have shown that contact angle can be tuned by varying solid roughness in the hydrophilic region ($\theta < 90°$). Other groups (Lenz and Lipowdky 1998; Li and Amirfazli 2007; Chen et al. 2007; Berthier 2008;

Hsieh et al. 2008) have shown that the substrate roughness amplifies the hydrophilic or hydrophobic character of contact.

Similar reasoning can be applied to a surface that is planar but chemically heterogeneous. The contact angle on chemically homogeneous and nonhomogeneous surfaces is shown in Fig. 1.5b. Viewing a chemically heterogeneous surface, as composed of distinct patches (such as A_1 and A_2) of various species, the apparent contact angle follows the relation

$$\cos \theta^* = A_1 \cdot \cos \theta_1 + A_2 \cdot \cos \theta_2 \tag{1.3}$$

Equation (1.3) is called the Cassie-Baxter relation; θ^* is the apparent contact angle, θ_1 and θ_2 are the local contact angles for surface patches 1 and 2 respectively, A_1 and A_2 are the fractional areas occupied by surface patches 1 and 2, respectively. Therefore, the apparent angle θ^* (restricted to the interval $[\theta_1, \theta_2]$) is given by an average involving the cosines of the angles characteristic of each constituent specie.

This discussion clarifies why the three-phase contact line of a liquid drop resting on a surface gets locally deformed: chemical and topographical heterogeneities play an important role. Certain surfaces have roughness in the form of micro-pillars creating a superhydrophobic substrate. Here, it has been observed that the drop can sit on a textured (rough) surface in two distinct configurations, Fig. 1.6a-i. For the Wenzel state, drop penetrates the pillars. For Cassie state, it does not contact the actual surface and indeed, may stay on the top of the pillars, Fig. 1.6a-ii. In such cases the contact angle is obtained as follows. For Wenzel state, (1.2) (Wenzel's law) applies as

$$\cos \theta_w = f \cdot \cos \theta \tag{1.4}$$

Here, θ_w is the apparent contact angle of the Wenzel state and f is the equivalent roughness of the substrate. For the Cassie state, one can write Cassie's law (1.3) as

$$\cos \theta_c = f \cdot \cos \theta + (1 - f) \cdot \cos \theta_0 \tag{1.5}$$

Here, θ_c is the apparent contact angle of Cassie state of drop and θ_0 is the contact angle with the layer of air, and f is the ratio of the contact surface (top of the pillars) to the total horizontal surface. If the pillars are not too far from each other, the value of $\theta_0 = \pi$.

1.5.2   Pinning of the Contact Line

In Fig. 1.6c, a sketch of the wetting behavior of a drop of liquid on a substrate with a continuously varying topography and continuously varying wettability, is depicted. Ondarçuhu (1995) and Lenz and Lipowdky (1998) showed the three-phase contact line to be pinned at a surface defect and a sharp transition of wettability on the

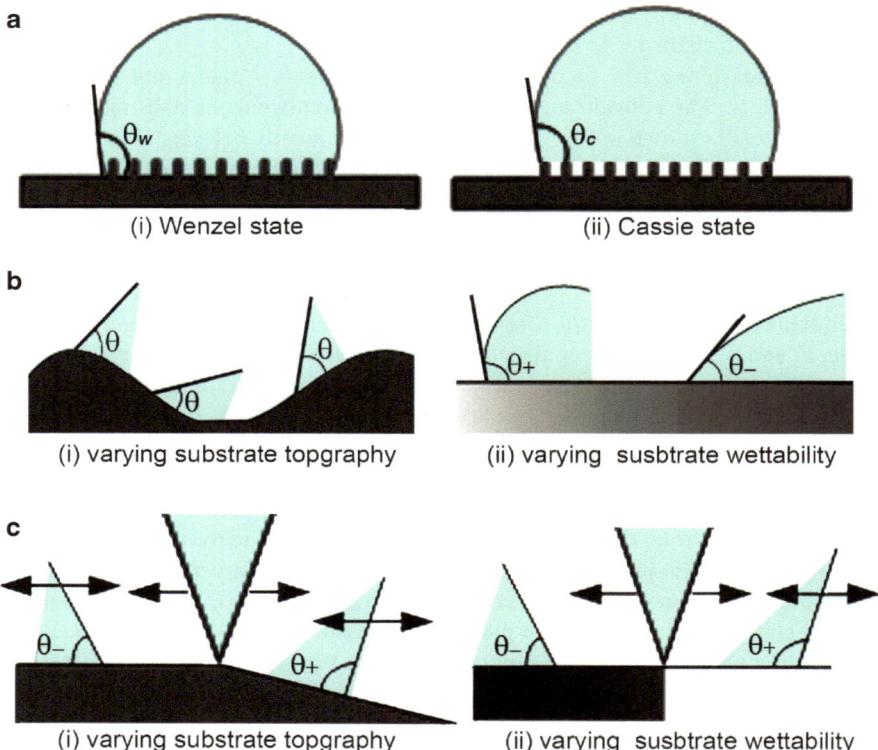

Fig. 1.6 Effect of contact angle when droplet is sitting on a textured substrate. (**a**) Two possible states of a drop sitting on a physically textured surface. (**b**) Effect of roughness and continuously varying wettability of substrate on contact angle. (**c**) Pinning due to sudden physical and chemical discontinuity

substrate. Local chemical and geometrical defects locally modify the contact angle. If the drop size is smaller than the length scale of the topography, Fig. 1.6c-i shows that the drop shape is not affected by the topography. If the drop is larger than the topographical features, the global shape of the drop will be affected by the deformation of the three-phase contact line. Similarly, for a substrate with a gradient in wettability (chemically nonhomogeneous surface), Fig. 1.6c-ii, the drop is deformed due to the peripheral changes in the contact angle of the three-phase contact line.

If the substrate has a sharp topography or wettability pattern, the situation is different, Fig. 1.6c-i. At the discontinuities, Young's Equation (1.1) becomes ill-defined. As a result, the three-phase contact line becomes immobilized. This effect is known as the pinning of the contact line. The pinning of an advancing contact line towards a convex edge over the substrate with homogeneous wettability is illustrated in Fig. 1.6c-ii. The contact angle at the boundary can have any value in between the smaller angle θ^- on the hydrophilic part and the larger value

θ^+ on the hydrophobic part. As a consequence, the position of the contact line is fixed to the line of discontinuity as long as the contact angle falls in the range of θ^- to θ^+. The contact angle now depends on the local wettability of the substrate and the global shape of the liquid–vapor interface at equilibrium. Contact angles will change further under dynamic conditions when, owing to fluid motion, a nonuniform pressure field is created within the drop. Hence, the contact angle is not only governed by the local wettability of the substrate but also depends on global shapes of the liquid–vapor interface in equilibrium. The wettability pattern on or underneath the substrate may act as an anchoring point for the contact line of a wetting liquid.

1.5.3   Capillary Length Scale

Surface tension, a negligible weak force in the macroscopic world, is dominant at smaller scales. This is because the force due to surface tension decreases linearly with size whereas weight scales down as the third power (Trimmer 1989). The crossover occurs at around the capillary length. Well below this crossover, the force due to surface tension is dominant and well above, the force of gravity is important.

To determine the capillary length, consider a liquid droplet underneath a substrate. It hangs underneath the substrate due to surface tension. It is stable until it grows large enough to be separated by the force due to gravity, i.e., its own weight. The force due to surface tension is approximated as

$$F_\sigma \approx r\sigma \tag{1.6}$$

The force due to the gravity is approximated as

$$F_g \approx r^3 \rho g \tag{1.7}$$

The two forces are equal when the drop separates from the substrate. The critical radius of drop, when it separates is obtained by balancing (1.6) and (1.7) and it equals the capillary length

$$l_c = r_c = \sqrt{\sigma/\rho g} \tag{1.8}$$

Capillary length defines the length scale below which surface tension dominates gravity. It is in the millimeter range for water. It depends on thermophysical properties of the liquid–solid combination. Leach et al. (2006) and Leipertz (2010) reported that small drops are locations of high heat transfer rates. Heat transfer diminishes with increasing drop radius. The largest drop diameter depends the interfacial forces at phase boundaries and body force, and hence, capillary length. Therefore, heat transfer coefficient in dropwise condensation crucially depends on capillary length or capillary radius.

Fig. 1.7 (**a**) Basic definition of advancing (θ_{adv}) and receding (θ_{rcd}) angles during immersion and removal of a solid plane in a liquid medium. (**b**) Droplet angle on an inclined plane with leading side angle (θ_{max}) and rear side angle (θ_{min})

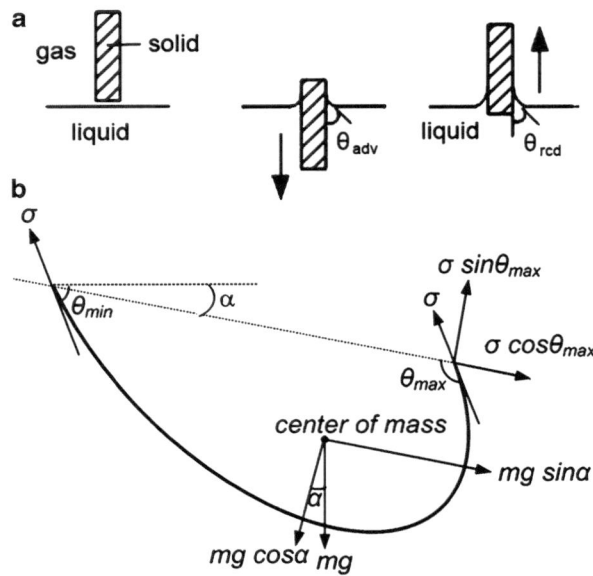

1.5.4 Contact Angle Hysteresis

Partially immersing a thin solid sheet in a liquid and moving it slowly, Furmidge (1962) reported the appearance of two distinct contact angles. These angles are known as the advancing angle θ_{adv} and receding angle θ_{rcd}, depending on the direction of motion of the plate, Fig. 1.7a. Arising from this experiment, a difference between the advancing and receding contact angles is known as contact angle hysteresis. For an idealized solid surface that is perfectly smooth, clean, and homogeneous in composition there would appear to be no reason for θ_{adv} and θ_{rcd} to be different. However, such an idealized surface does not exist. Real condensing surfaces are typically metallic and are never perfectly smooth, composition may vary slightly with location, and molecules, atoms, or ions of other substances may be adsorbed on surface. Contact angle hysteresis is acknowledged to be a consequence of three factors: (1) surface inhomogeneity, (2) surface roughness, and (3) impurities on the surface.

If a drop of liquid is placed underneath a horizontal surface, it achieves an equilibrium shape and leaves the droplet with almost constant angle all around its perimeter (ideal contact angle hysteresis is zero). If the surface is then turned through angle α, the drop will deform to balance the gravity force parallel to plate inclination and perpendicular to the substrate, Fig. 1.7b. The contact angle reaches its maximum value for an advancing liquid edge of the drop and the minimum value for a receding liquid edge of the drop. Many researchers (Brown et al. 1980; Lawal and Brown 1982; Extrand and Kumagai 1995; ElSherbini and Jacobi 2004a, b; Dimitrakopoulos and Higdon 1999) argued that the maximum and minimum contact angles on or underneath an inclined substrate are equivalent to advancing and receding contact angles respectively. The static advancing and receding contact angles can be easily observed from a tilted pendant droplet as shown in Fig. 1.7b. In a dynamic context, the contact angle changes from static values so that the advancing angle increases and receding angle decreases as a function of the speed of the three-phase contact line. The literature on the interrelation between the static and dynamic contact on or underneath an inclined substrate is rather scarce.

Contact angle hysteresis plays an important role in the stability of a drop on or underneath an inclined substrate. The difference between the top and bottom side contact angle makes it possible for the droplet to adopt a shape that may support the weight of the liquid against gravity. Hence, contact angle hysteresis offers resistance against the motion of drop. Note in Fig. 1.7b that the interface radius of curvature is smaller over the upper portion of the droplet and greatest near the base. The resulting variation of the surface tension forces over the interface may serve to balance the hydrostatic pressure difference across the interface, allowing the droplet to hold its position on the vertical wall against gravity.

Various investigators (Dussan 1979; Dussan and Chow 1983; Dussan 1985; Briscoe and Galvin 1991a, b; Extrand and Kumagai 1995; Miwa et al. 2000; Elsherbini and Jacobi 2006) have derived an expression relating the criticality of drop sliding on or underneath an inclined plane in dropwise condensation, as a function of hysteresis ($\theta_{adv} - \theta_{rcd}$), as follows

$$\sin \alpha = \sigma_{lv} \cdot (R \cdot K/m \cdot g)(\cos \theta_{rcd} - \cos \theta_{adv}) \qquad (1.9)$$

In (1.9), α is the critical sliding angle, σ_{lv} the surface tension, m the mass of the drop, and R and K are a length scale and shape constant for the contour of the drop, respectively.

Öner and McCarthy (2000) made it clear that contact angle hysteresis can be a qualitative indication of drop mobility. Yet, Krasovitski and Marmur (2005) and Pierce et al. (2008) argued that advancing and receding contact angles are measured on a level surface and should theoretically not be used in numerical predictions of the sliding angles. Instead, they define the maximum and minimum contact angles (θ_{max} and θ_{min}), which are those that occur at the leading and trailing edges of a drop profile on a surface inclined at the sliding angle, Fig. 1.7b. The modified form of (1.9) is obtained as follows

$$\sin \alpha = \sigma_{lv} \cdot (R \cdot K/m \cdot g)(\cos \theta_{max} - \cos \theta_{min}) \qquad (1.10)$$

Theoretical and experimental evidence suggest that the relationship between θ_{max} and θ_{adv} as well as θ_{min} and θ_{rcd} respectively, varies with the surface-liquid combination. ElSherbini and Jacobi (2004a, b) reported empirical data that exhibits θ_{max} and θ_{min} approximately equal to θ_{adv} and θ_{rcd} for all surface-water combinations. Krasovitski and Marmur (2005) reported that the upper side contact angle (θ_{min}) tends to be approximately equal to the receding contact angle, while the lower side contact angle (θ_{max}) may be much lower than the advancing contact angle. Hence, there is some controversy on the value of the leading angle and trailing angle of a deformed drop on an inclined substrate at criticality. This information is quite important from the viewpoint of dropwise condensation and has attracted attention. In the present monograph, the leading side angle (θ_{max}) is assumed equal to the advancing angle (θ_{adv}) of drop and trailing side angle (θ_{min}) equal to the receding angle (θ_{rcd}) at criticality for determining the size of drop at criticality.

Various researchers (Leach et al. 2006; Ma et al. 2008; Kim and Kim 2011; Rykaczewski 2012) have reported that heat transfer rate increases with diminishing contact angle hysteresis since criticality of drop slide-off/fall-off is inversely proportional to it. Large hysteresis will provide adequate forces along and normal to the wall and improve the stability of the drop. Conversely, the drop slide-off or fall-off will occur early on a surface that has small hysteresis. The repeated sweep and removal of drops from a surface result in fresh condensation and an overall improvement in the heat transfer rate.

On a roughened hydrophobic substrate, a liquid drop can exhibit either the Cassie state where the drop sits on the air-filled textures or the Wenzel state where the drop wets cavities of the textures (see Fig. 1.6). The apparent contact angle of a roughened hydrophobic surface is enhanced in both the Cassie and Wenzel states; however, the Cassie state is the preferred superhydrophobic state in which a drop has a much smaller contact angle hysteresis and therefore a higher mobility. Till date, none of the reported condensation studies on engineered superhydrophobic surfaces has exhibited a sustained Cassie state; instead, the condensate drops partially or fully penetrates into the cavities over the course of condensation (Ma et al. 2012).

1.5.5 Vapor Accommodation Coefficient

When condensation occurs, kinetic theory of gases suggests that the flux of vapor molecules joining the liquid must exceed the flux of molecules escaping the liquid phase. Accommodation coefficient, denoted as $\hat{\sigma}$ defines the fraction of the striking vapor molecules that actually get condensed on the vapor–liquid interface. The remaining fraction $(1 - \hat{\sigma})$ is due to reflection of vapor molecules that strike the interface but do not condense. The accommodation coefficient indirectly measures the interfacial resistance of the liquid–vapor interface to condensation. Higher the accommodation coefficient, lower the interfacial resistance of the liquid–vapor interface of the condensed drop. Quoted values of $\hat{\sigma}$ in literature widely vary. Mills and Seban (1967) reported that the accommodation coefficient is less than unity only when the interface is impure. For pure liquid–vapor interface, the value reported in the literature is unity. Because extreme purity is unlikely in most engineering systems, a value of less than unity can be expected. Sukhatme and Rohsenow (1966) reported its values ranging from 0.37 to 0.61 for condensation of metallic vapor. For liquid ethanol, methanol, alcohol, and water, the reported values

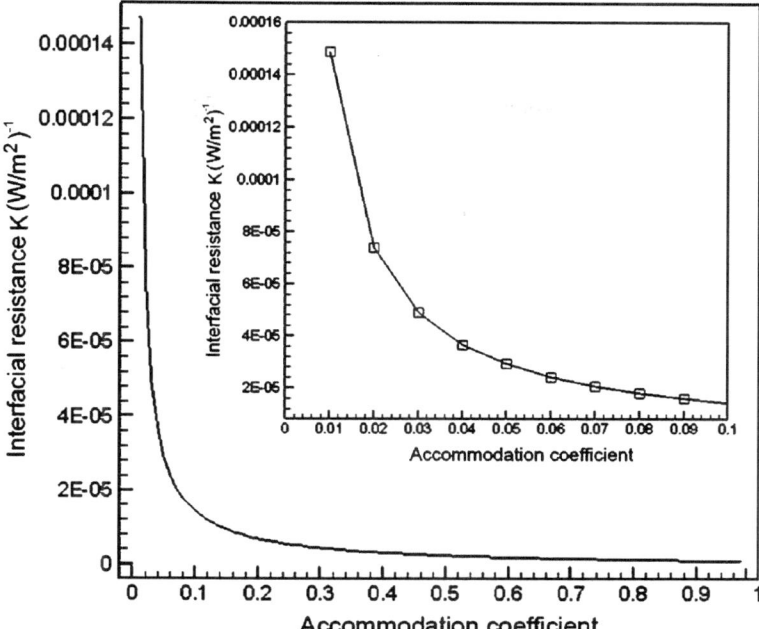

Fig. 1.8 Variation of interfacial heat transfer coefficient with respect to the accommodation coefficient for water vapor

of accommodation coefficient range from 0.02 to 0.04 (Carey 2008). On the other hand, for benzene and carbon tetrachloride, reported values are closer to unity. Mareka and Straub (2001) reported that accommodation coefficient decreases with increasing temperature. The interfacial resistance may be particularly important in the condensation of liquid metals.

The variation of interfacial heat transfer resistance per unit area (see Chap. 2), at experimental conditions of 30 °C saturation temperature and 0.015 bar saturation pressure for water vapor, are presented in Fig. 1.8. There is considerable variation of the interface heat transfer coefficient of over an order of magnitude for small values of the accommodation coefficient, $0.01 < \hat{\sigma} < 0.1$. Beyond $\hat{\sigma} > 0.1$ the interface heat transfer coefficient does not change appreciably.

1.6 Closure

This chapter introduces briefly the classification and significance of various physical processes in dropwise condensation. The importance of wettability and contact angle on the formation of drops is highlighted. The shape of the drop plays a central role in fixing conduction resistance, the onset of instability with respect to static equilibrium, as well as its motion over the substrate. Once large drops move out of the surface, fresh nucleation ensures that the condensation process is cyclic, with a characteristic timescale, area coverage, and drop size distribution. Mathematical modeling of the dropwise condensation process forms the topic of the following chapters.

Chapter 2
Modeling Dropwise Condensation

Keywords Mathematical modeling • Nucleation • Population balance • Minimum radius • Growth by direct condensation • Coalescence • Instability • Condensation cycle

2.1 Mechanism of Dropwise Condensation and its Modeling

The large body of literature available on the subject suggests the following three independent mechanisms of dropwise condensation (Leipertz 2010):

1. The vapor condenses primarily between the droplets, i.e., the droplet-free area. This condensate layer gets transported to the droplets in their vicinity by surface diffusion. According to this model, the thin film between the droplets and the free surface of the droplets contribute to overall heat transfer.
2. While vapor condensation begins in a filmwise mode (filmwise condensation), the film reaches a critical thickness and ruptures due to surface tension driven instability forming droplets. It is postulated that major part of the heat transfer takes place at this very thin condensate film, while the droplets mainly act as liquid collectors. This model of the dropwise condensation process was proposed by Jakob (1936). Song et al. (1991) have put forward a droplet and condensate film mechanism for the formation of droplets during dropwise condensation. These authors observed that a thin film of condensate exists on open areas among the droplets and a film of condensate remains at the spots from which the droplets have departed.
3. Droplets are only formed at individual nucleation sites, while the area between the droplets is regarded to be inactive with respect to condensation. In this model, heat transfer occurs only through the droplets and is primarily limited by their heat conduction resistance. This model was first proposed by Eucken (1937). Majority of the studies support this mechanism, in which the condensate is in the form of discrete drops located at the nucleation sites on or underneath a lyophobic substrate.

S. Khandekar and K. Muralidhar, *Dropwise Condensation on Inclined*   17
Textured Surfaces, SpringerBriefs in Applied Sciences and Technology 11,
DOI 10.1007/978-1-4614-8447-9_2, © Springer Science+Business Media New York 2014

McCormic and Baer (1963) proposed a new mechanism of heat transfer in dropwise condensation. The analysis indicated that heat is transferred through active areas on the condenser surfaces which are continually produced by numerous drop coalescence. These areas remain active for a short portion of the cycle time. During this time, numerous submicroscopic drops grow from randomly distributed sites. McCormick and Westwater (1965) studied nucleation of water drops during dropwise condensation on a horizontal surface of copper coated with a monolayer of benzyl mercaptan. Their photographic evidence showed no visible condensate liquid film among the droplets. Drops nucleated at natural cavities on the condenser surface. Some cavities were nucleation sites because they contained trapped, liquid water.

Umur and Griffith (1965) found that, at least for low temperature difference, the area between growing droplets on the surface was, in fact dry. Their results indicate that no film greater than monolayer thickness existes between the droplets, and no condensation can take place in these areas. Further evidence of nonexistence of a condensate film between droplets was furnished by Ivanovskii et al. (1967), using a different fluid. By measuring the electrical resistance between the two electrodes embedded in a glass surface on which dropwise condensation of mercury was taking place, they concluded that no thin condensate layer existed between droplets.

Photographs taken through a microscopic with magnification of up to 400 showed nucleating and growing droplets which eventually coalesce with neighboring droplets (McCormick and Westwater 1965). New drops form on the sites vacated by the coalescing droplets.

The first dropwise condensation model was proposed by Le Fevre and Rose (1966). In this model, a calculation for heat transfer through a single drop was combined with that of the drop size distribution to obtain the average heat flux. For deriving heat transfer rates through a single drop, the following three thermal resistances are considered: (a) conduction resistance, (b) vapor–liquid interfacial resistance, and (c) surface curvature resistance. The thickness of the promoter layer was neglected.

Gose et al. (1967) developed a model for heat transfer during dropwise condensation on randomly distributed nucleation sites. Simulation was performed on a 100×100 grid with 200 randomly distributed nucleation sites. The model accounted for growth, coalescence, vacating active sites beneath the smaller of the coalescing drops, re-nucleation on the newly exposed sites, and drop removal. For steady-state condensation, the theory showed that small drops grow by vapor condensation, and that larger drops grow predominantly by coalescence. The authors observed that higher nucleation sites and drop removal from substrate were factors for a large heat transfer coefficient.

Glicksman and Hunt (1972) simulated the condensation cycle in a number of stages, covering the equilibrium drop size to the departing drop size, with a large nucleation site density. The initial nucleation site density considered was $10^5 \, \mathrm{cm}^{-2}$ with 1,000 sites on a surface of size 33 μm $\times$ 33 μm. The area of the second stage was increased ten times and the droplets from the first stage were redistributed on this surface. In this way, the simulation was repeated until the departure droplet size was reached. An artificial redistribution between the two consecutive stages, however, destroyed the natural distribution of drops.

Rose and Glicksman (1973) proposed a universal form of the distribution function for large drops which grow primarily by coalescence with smaller drops, though smaller drops themselves grow by direct condensation.

Tanaka (1975a, b) used a precise expression for the calculation of drop size distribution. The author considered the transient change of local drop size distribution, taking into account the processes of growth and coalescence of drops. From this point of view, the author put forward a theory of dropwise condensation. The theory is based on the following assumptions: (1) primary droplets nucleate at discrete sites distributed randomly on the condensing surface; (2) drops are hemispherical; (3) the governing heat-transfer resistance through a single drop is heat conduction; and (4) temperature of the condensing surface is uniform. Basic integro-differential Equations describing the transient process of dropwise condensation on a newly swept region were derived. By introducing a model for the cycle of drop departure, a general expression for the average heat-transfer coefficient was obtained.

Wu and Maa (1976) used the population balance method to find the drop size distribution of small drops which grow mainly by direct condensation. They estimated the heat transfer coefficient by considering the conduction resistance through the drop. Maa (1978) later utilized the population balance equation derived for dropwise condensation, considering both drop growth due to direct condensation and coalescence between drops, to obtain the resulting drop size distributions. Results confirmed that the drop size distribution and heat flux of dropwise condensation depend strongly on the concentration of active nucleation sites on the substrate surface.

Meakin (1992) described the following stages of the dropwise condensation: (a) nucleation and growth, (b) growth and coalescence, (c) growth and coalescence with renucleation in exposed regions, and (d) growth, coalescence, and renucleation with removal of larger droplets. All the four stages were simulated and their results were described in terms of simple scaling theories.

Abu-Orabi (1998) used the population balance approach to predict the drop size distribution for small drops that grow by direct condensation. Resistances to heat transfer due to the drop (conduction through the drop, vapor–liquid interfacial resistance, and drop curvature), the promoter layer, and the sweeping effect of the falling drops were incorporated into the model. The total heat flux was calculated from the drop size distributions and the heat transfer rate through a single drop. Drop size distribution for large drops that grow by coalescence was obtained from the work of Rose and Glicksman (1973).

Burnside and Hadi (1999) simulated dropwise condensation of steam from an equilibrium droplet to a detectable size on 240 μm $\times$ 240 μm surface with 10^8 cm^{-2} randomly spaced nucleation sites, stopping when the maximum drop radius was about 4 μm. The authors observed a maximum drop radius of 3.9 mm, 0.21 ms after the start of condensation and peak heat transfer coefficient immediately after the condensing surface wipe up by the drop.

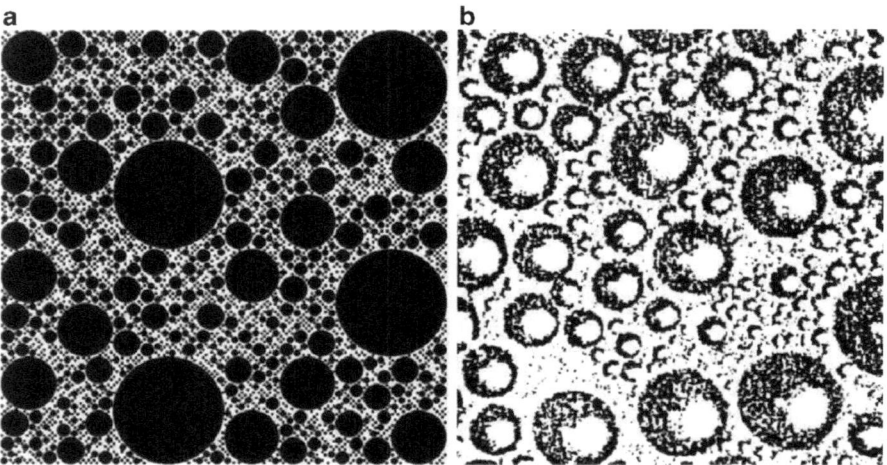

Fig. 2.1 Comparison of drop distribution between random fractal model and direct photography (Wu et al. 2001)

Wu et al. (2001) presented a fractal model to simulate drop size and its spatial distribution in dropwise condensation. The boundary conditions of heat conduction through the condensing surface were established using the heat transfer model through a single drop proposed by Rose (1981). Photographs of dropwise condensation at various instants were similar to experiments, Fig. 2.1.

Vemuri and Kim (2006) modeled dropwise condensation for hemispherical drops which grow by direct condensation, using the population balance method. The primary resistances to heat transfer, such as conduction through the drop and vapor–liquid interface were considered. The derivation of steady state distribution for small drops within the size range of negligible coalescence was based on the conservation of the number of drops with no accumulation. Contact angles other than 90° were not considered in this model.

On growth kinetics, Leach et al. (2006) reported from experiments that the smallest drops grow principally by the accretion of liquid molecules diffusing along the substrate surface, while drops larger than about 50 μm in diameter grow by the deposition of condensing vapor directly onto the drop surface. The effects of contact angle, degree of subcooling, and inclination of substrate for a hydrophobic polymer film and silanized glass surface for sessile droplets were reported.

Liu et al. (2007) showed experimentally proved that the state of initial condensate formed on the surfaces is not in the form of a thin film but as nuclei. These results demonstrate that the mechanism of formation of initial condensate drops for dropwise condensation accords with the hypothesis of nucleation sites. Consequently, recent analytical models have assumed that droplets form on nucleation sites, neglecting any heat transfer taking place between the drops. It is also assumed that condensation occurs only on the free surface of droplets, and that the latent heat is transferred through the droplets to the solid surface.

Similarly, experimental and theoretical work of Carey (2008) casts serious doubt on the existence of films and supports the view of McCormic and Baer (1963) that nucleation is an essential feature of dropwise condensation.

Kim and Kim (2011) modeled dropwise condensation over a superhydrophobic surface. The overall methodology, similar to those described earlier, has the following differences: (a) Heat transfer through a single droplet is analyzed as a combination of the vapor–liquid interfacial resistance, the resistance due to the conduction through the drop itself, the resistance from the coating layer, and the resistance due to curvature, (b) Population balance model is adapted to develop a drop distribution function for the small drops that grow by direct condensation, (c) Drop size distribution for large drops that grow mainly by coalescence is obtained from the empirical equation of Tanaka (1975b). Results showed that the single droplet heat transfer and drop distribution are significantly affected by the contact angle.

A complete simulation of dropwise condensation from drop formation at the atomic scale to the departing droplet size, accounting for the effect of saturation temperature, contact angle, and contact angle hysteresis, wettability gradient on the condensing substrate and the inclination of the substrate along with its experimental validation are reported in the present work. The model presented in this work is based on the postulation that drop embryos form and grow at nucleation sites, while the portion of the surface between the growing drops remains dry. The vapor condenses on the free surface of drops at each of the nucleation sites. Latent heat released during condensation is transferred through the liquid drop to the cold wall. Thus, heat transfer in dropwise condensation is primarily limited by the thermal resistance of the liquid drop and the available nucleation site density.

Dropwise condensation is a combination of various processes occurring over a wide spectrum of length and timescales. A comprehensive mathematical model of various subprocesses occurring in dropwise condensation underneath an inclined textured substrate is reported in this chapter.

A framework that explains hierarchical modeling of dropwise condensation in terms of the processes involved is depicted in Fig. 2.2. The atomistic model, which relies on population balance, is the starting point for determining the size of the smallest stable drop. The nucleation sites are randomly distributed on the substrate and all the sites are initially occupied by the drops of smallest radius—namely, the maximum size of a stable cluster) in the atomistic model. The subsequent steps that follow are growth by direct condensation, coalescence, instability, drop motion, and computation of transport coefficients for sliding drops. The model as a whole yields the instantaneous drop size distribution, instantaneous rate of growth of drops, area of coverage by drops, frequency of drop slide/fall-off, and local and average heat transfer coefficient over inclined surfaces. The simulation discussed here is confined to condensation underneath cold inclined substrates forming pendant drops.

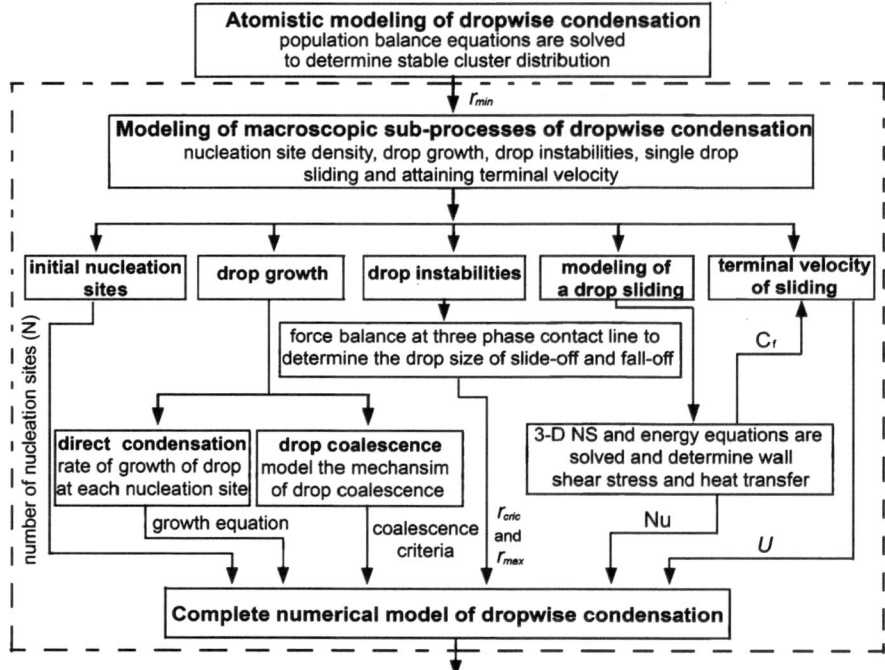

Fig. 2.2 Schematic diagram of hierarchical modeling in dropwise condensation

2.2   Drop Formation at the Atomic Scale

Condensation in the form of discrete drops can be homogeneous, namely, distributed in the vapor phase, or heterogeneous, as in the presence of a cooler solid substrate. It is now accepted that phase-change, whether homogeneous or heterogeneous, is induced by nucleation, triggered by molecular clustering. In view of experimental limitations, the physical picture, right at nucleation, is not very clear. From a heat transfer viewpoint, an important question is how do drops form, grow, and get mobilized over a textured solid surface. At the atomic level, vapor atoms may impinge on the surface with a direct velocity, or alternatively, the vapor may be quiescent. Individual quiescent vapor atoms may form stable clusters by combining with neighbors and grow on the substrate with time by losing their latent heat (Sikarwar et al. 2012).

Bentley and Hands (1978) reported various processes occurring at the atomic scale, Fig. 2.3, from arrival of monomers to formation of stable clusters on the cold substrate. The surface adatoms undergo a sequence of processes such as adsorption, diffusion, reflection, agglomeration, transfer of energy, and formation of stable clusters, eventually manifesting as a distribution of condensed liquid nuclei.

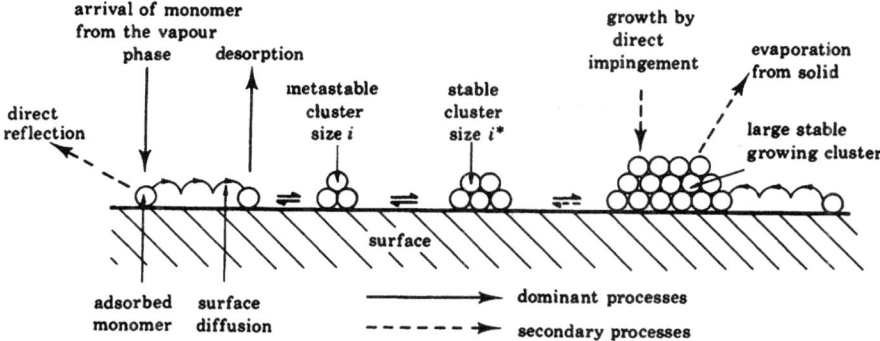

Fig. 2.3 Processes involved in deposition of condensate atoms on a cold substrate (Bentley and Hands 1978)

Atoms/molecules bound to the surface form an adatom and a group of adatoms leads to a cluster. Although it is possible to form clusters in the vapor phase before they get deposited on the surface, with large substrate subcooling, one can expect all the condensation to occur at the surface leading to heterogeneous condensation.

Lee and Maa (1991) experimentally observed the mechanism of vapor deposition by using an electron microscope. The process was composed of adsorption of the vapor molecules on the substrate, surface diffusion, growth, and coalescence of the deposited clusters.

Hashimoto and Kotake (1995) reported that molecules approaching the cooled wall have higher energies than the departing molecules that have transferred their energy to the wall. The energy exchange between the incoming and departing molecules at distinct temperatures provides a sufficient condition for molecular clustering, Fig. 2.4. The cluster size formed increases near the wall, and the thickness of the cluster zone depends on the thermal condition of the molecular system and the processes of energy transfer.

Kotake (1998) reported that the existence of clusters depends on the condition of energy transfer between molecules of the vapor to the cold substrate. The rate of condensation depends on the relative strengths of three intermolecular attractions: (a) the energy of attraction between two adsorbed molecules, (b) the adsorption energy on the substrate, and (c) the adsorption energy of a vapor molecule on an adsorbed layer of its own species.

Peng et al. (2000) reported that clustering of molecules on a cold substrate is similar to reaction kinetics with appropriate reaction rates. The authors reported that the driving force for the coalescence of clusters results from the tendency of minimization of surface energy. The transport of mass occurs via the routes of evaporation–condensation and surface diffusion.

McCoy (2000) presented a theory based on cluster distribution kinetics for single monomer addition and dissociation. Population balance equation was used to describe the dynamics of cluster mass distribution during homogeneous and heterogeneous nucleation in unsteady closed and steady flow systems. The distribution-kinetics

Fig. 2.4 Cluster formation in the vicinity of the condensate or a cooled wall (Kotake 1998): evaporating molecules have lower energy (e_e) than condensing molecules (e_c); $e_e < e_c$ is a favorable condition for cluster formation

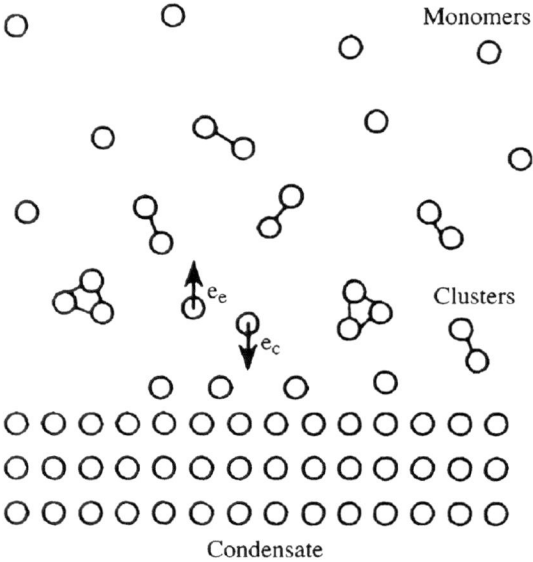

approach was based on the recognition that nucleation and growth from vapor led to droplets larger than the nuclei and distributed in mass. Cluster growth by addition–dissociation was found to be similar to polymerization–depolymerization reaction. It was shown that heterogeneous nucleation preserves the number of clusters, equal to the nucleation sites.

Wang et al. (2003) proposed an idea of critical aggregation concentration of active molecules to describe the moment just before nucleus formation. Tian et al. (2004) studied the aggregation of active molecules inside a metastable bulk-phase using thermodynamics. The authors derived an expression for the critical aggregation concentration, energy distribution of active molecules inside the bulk-phase at superheated and supercooled limits, and used the molecular aggregation theory to describe the gas–liquid phase-transition process.

Song et al. (2009) suggested that steam molecules become clusters prior to condensation on the cooled surface, Fig. 2.5. The authors argued that clustering begins in the vapor phase itself close to the cold wall. Clusters formed closer to the wall are larger than those formed in the bulk vapor phase.

On the basis of the available literature, it can be concluded that drop formation during condensation commences with the impingement of vapor atoms on or underneath a cold substrate. Alternatively, vapor may be quiescent. The individual quiescent atoms may form stable clusters by combining with the neighbors and grow on the surface with time, Fig. 2.6a. An atom/molecule bound to the surface is an adatom and a group of adatoms leads to a cluster, Fig. 2.6b. It is also possible to

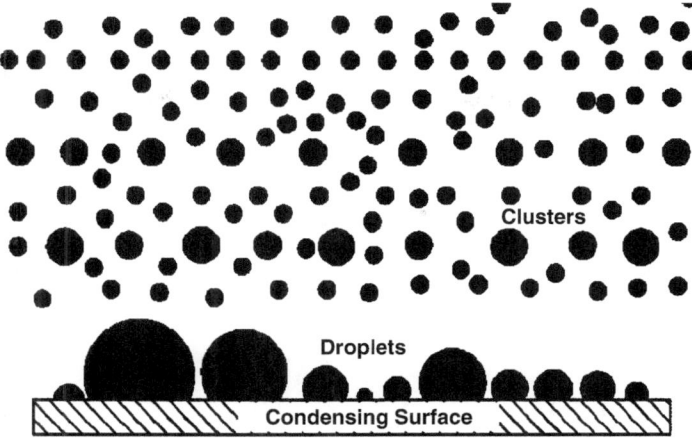

Fig. 2.5 Physical model of vapor condensation, proposed by Song et al. (2009)

form clusters in the vapor phase before they get deposited on the surface. With large substrate subcooling, one can expect all condensation to occur at the surface level. The stability of the cluster depends on mutual energy interactions between the cluster, the atoms of the surrounding vapor and the cold wall. Molecules/atoms approaching the cold wall have a higher temperature than departing molecules/ atoms that have transferred their energy to the wall. This energy difference determines whether a given cluster clinging to the surface will be stable, grow with time, or diminish in size. Many stable clusters growing together may form an atomic/molecular monolayer of condensate on the substrate, Fig. 2.6b, c.

There are at least two possibilities of drop formation (dewetting), Fig. 2.6c, d. In the first model, it is postulated that the condensation initially occurs in a filmwise manner, forming an extremely thin film on the solid surface, Fig. 2.6c-i. As condensation continues, this film ruptures due to intrinsic interfacial instabilities and distinct drops are formed, Fig. 2.6d-i. The second theory is based on the premise that drop formation is a heterogeneous nucleation process, Fig. 2.6c-ii. Here, stable clusters gets located at specific nucleation sites over the substrate, such as pits and grooves, grow in the continuum domain, while the portion of the surface between growing drops essentially remains dry. Small droplets are formed by direct condensation via nucleation at locations with local minima of the free surface energy, Fig. 2.6d-ii. Hence, the processes such as molecular potential, adatoms dynamics, cluster dynamics, surface diffusion, stable cluster size, nucleation density, film stability and rupture, topography interaction, and stable cluster formation appear as condensation proceeds from the atomistic scale to the microscale. From the heat transfer point of view, control of these processes is of critical concern in atomistic modeling of dropwise condensation.

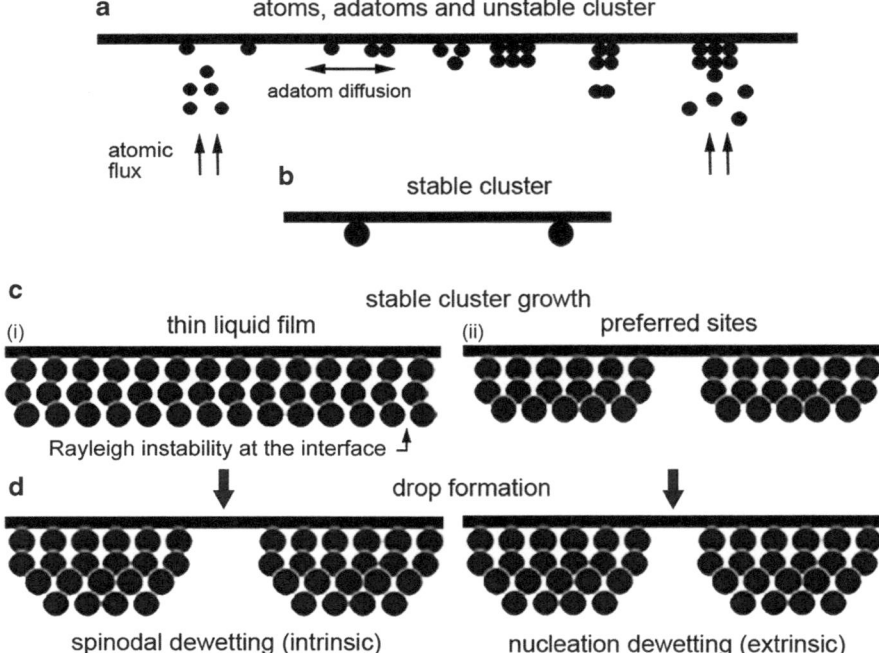

Fig. 2.6 Mechanism of liquid drop formation underneath a cold substrate. From angstroms to nanometers, individual vapor molecules come closer, a system of adatoms form and a group of adatoms leads to a cluster. Many growing clusters together may form a molecular monolayer of liquid. At this stage there are at least two possibilities: droplet formation (dewetting) and film formation. The liquid film ruptures and forms droplets

2.3 Atomistic Modeling of Dropwise Condensation

The basic aim of atomistic modeling of dropwise condensation is to determine the size of the stable cluster at nucleation and connect phenomena occurring at the atomic scale to the macroscale. Formation of drops during condensation commences with the impingement of vapor atoms on a cold substrate kept at a temperature below saturation. Atoms approaching the cold wall have higher energies than departing atoms and hence transfer energy to the wall. The energy exchange between incoming and departing atoms of different energies provides a sufficient condition for molecular clustering.

The vapor mass flux F is obtained in the form of an overexpanded jet from a nozzle discharging into an evacuated chamber, Fig. 2.7a. When the vapor is stationary, the mass flux is set to zero. The substrate on which all the condensation takes place is initially clean and free of any condensate. Atoms are deposited on the substrate at a constant rate. An adsorbed layer of atoms, called adatoms, is first formed prior to nucleation, Fig. 2.7b. These adatoms can diffuse on the surface with

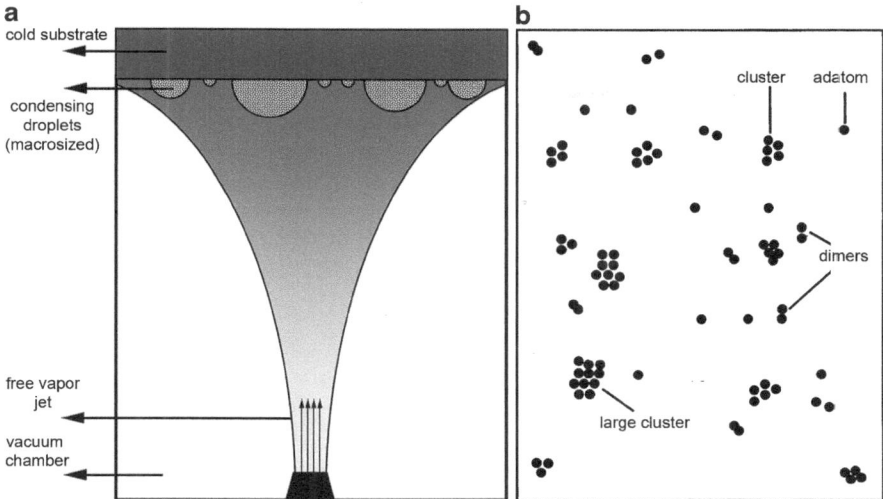

Fig. 2.7 Physical modeling of droplet formation underneath a substrate. (**a**) Schematic representation of the vapor flux impinging vertically on the underside of a horizontal substrate. (**b**) Schematic drawing of the distribution of clusters on the substrate

a characteristic time period that is the mean resident time (τ_{ads}) and then re-evaporate back to the vapor phase. They may collide with other adatoms or clusters during their migration, thus causing nucleation to be initiated. The adatom population on the substrate changes with time due to desorption, capture, or release of an adatom by a cluster. The population may redistribute itself over the surface as a result of diffusion at a speed determined by the diffusion coefficient. If two adatoms occupy neighboring sites, they will stick to form a cluster. More adatoms may be captured by a cluster or two clusters may combine to form large clusters. The population of cluster of a certain size will thus change due to adatom capture or release, coalescence with other clusters, or breakage into smaller clusters and desorption. In the growth stages, the condensate clusters grow, not only by capturing adatoms on the surface, but also by direct capture of impinging vapor molecules/atoms. The residence time is taken to be large enough so that sufficient time is available for all the adatoms existing in vapor phase to lose their latent heat and get condensed.

When the temperature of the substrate is significantly lower than the saturation temperature, condensation will be complete, in the sense that all the atoms contained in the vapor phase stick to the substrate. Under these conditions, the following additional assumptions facilitate the computation of cluster densities:

1. Adatoms alone diffuse while dimers and larger clusters are stable, namely, they do not disintegrate or diffuse within the substrate.
2. Direct impingement of free atoms on adatoms and clusters, and the coalescence of clusters can be neglected. Thus, the atoms and clusters diffusing within the substrate arise exclusively from the condensate and do not have contributions to their population from the vapor phase.

2.3.1 Mathematical Model

With the approximation discussed above, the rate equations of the atomistic model (Brune 1998; Amar et al. 2001; Venables 2000; Oura et al. 2003) reduce to:

$$(dn_1/dt) = F - (n_1/\tau_{ads}) + \left(-2\sigma_1 \cdot D \cdot n_1^2 - n_1 \cdot \sigma_x \cdot D \cdot n_x\right) \qquad (2.1)$$

$$\left(dn_j/dt\right) = n_1 \cdot \sigma_{j-1} \cdot D \cdot n_{j-1} - n_1 \cdot \sigma_j \cdot D \cdot n_j \text{ for } j = 2 \text{ to } 1,000 \qquad (2.2)$$

$$(dn_x/dt) = n_1 \cdot \sigma_i \cdot D \cdot n_i \qquad (2.3)$$

Complete information on the local distribution of clusters is contained in the capture and decay rates, σ_j and δ_j respectively. In the present study, these quantities are given parameters. The capture coefficients are nearly constant with $\sigma_1 = 3$ and $\sigma_x = 7$. A first principles calculation of these parameters involves solving a Helmholtz-type diffusion equation for clusters in two dimensions in the presence of a certain density of stable islands. The analytical expressions obtained with this approach are (Brune 1998; Venables 2000):

$$\sigma_x = \frac{4\pi(1 - Z)}{\ln(1/z) - (3 - z)(1 - z)/2} \qquad (2.4)$$

$$\sigma_1 = 4\pi(1 - n_1)\frac{n_x}{n_1}\frac{1}{\ln(1/z) - (3 - z)(1 - z)/2} \qquad (2.5)$$

Here, $Z = \vartheta - \sum_{j=1}^{i} n_j$ is the fraction of the surface covered by the stable clusters and ϑ is the total coverage area. Using constant values of σ_1 and σ_x one can obtain the island size distribution for a specified value of i. For the present discussion, it is assumed that dimers as well as clusters with three or more atoms are stable; consequently the decay constants $\delta_j(j \geq 2)$ are effectively zero. The assumption is equivalent to stating that clusters that are held together by the long-range van der Waal forces do not have any intrinsic breakup mechanism. The long-range forces appear over length scales of a few nanometers while repulsive forces become significant over considerably shorter length scales of a few angstroms. Thus, number densities of clusters change purely because of the addition of monomers.

The condition that complete condensation of the impinging vapor takes place is equivalent to the inequality $\sigma_x \cdot n_x \cdot D \cdot \tau_{ads} >> 1$; it neglects re-evaporation effects (Venables 2000). For the complete condensation regime modeled here, the mean residence time τ_{ads} is high. It was found that the model predictions reported in the present study were not sensitive to changes in this quantity for $\tau_{ads} \geq 0.1$ seconds.

2.3.2 Numerical Methodology

Numerical simulation of (2.1–2.3) was run for a large set of cluster sizes varying from adatoms (cluster containing one atom/molecule) to clusters containing 1,000 atoms/molecules. The largest cluster with a nonzero number density was found from simulation to have 100–200 atoms/molecules. Hence, the choice of a cluster with 1,000 atoms as an upper limit was considered adequate.

The initial conditions were specified during simulation as $n_j(t = 0) = 0$ (for $j = 1$ to 1,000) and $n_x(t = 0) = 0$. The model parameters were taken as $\sigma_1 = 3$ (for $j = 1$), $\sigma_j = 7$ (for $j = 2$ to $1,000$) and $\sigma_x = 7$.

Brune (1998) has showed that the values of the capture coefficients specified above give meaningful results; the corresponding computational effort is also lower since they need not be repeatedly calculated from (2.4 and 2.5). A vapor flux of $F = 0.005$ per second has been adopted for the study. The diffusion constant D was calculated with the ratio (D/F) taking on the values of 10^5, 10^6, 10^7. The residence time of $\tau_{ads} = 2.3$ s was chosen from numerical experiments to model the complete nucleation regime.

Equations (2.1–2.3) constitute a system of 1,001 coupled ODEs. The fourth order Runge-Kutta method was implemented in a C++ language program to solve the system of simultaneous differential equations. The model and the computer program were validated against the benchmark results and are presented next.

2.3.3 Validation

The validation of cluster growth simulation is discussed here. The number of islands/clusters of size s can be expressed in terms of the scaling function (Bartelt et al. 1993; Ratsch and Zangwill 1994; Stroscio and Pierce 1994; Brune et al. 1999; Ratsch and Venables 2003; Shi et al. 2005).

$$n_s(\vartheta) = \frac{\vartheta \cdot f_i(s/S)}{S^2} \tag{2.6}$$

The symbol n_s is the number of islands of size s at coverage ϑ, given by:

$$\vartheta = \sum_{s \geq 1} s \cdot n_s \tag{2.7}$$

The average island size is

$$S = \sum s \cdot n_s \Big/ \sum n_s \tag{2.8}$$

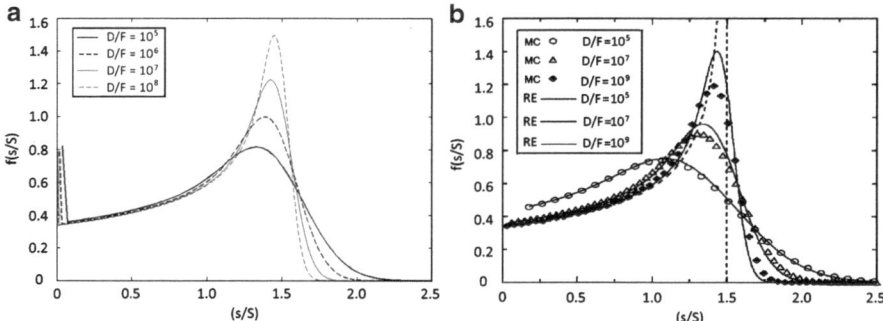

Fig. 2.8 (a) Numerical simulation of the rate equations governing the nucleation process. (b) The results given by Shi et al. (2005) by using both the rate equations (RE) approach (*solid lines*) and Monte Carlo (MC) simulation (*symbols*) are presented here

Fig. 2.9 Variation of monomer density (n_1) and saturation island density (n_x) with coverage at various values of D/F

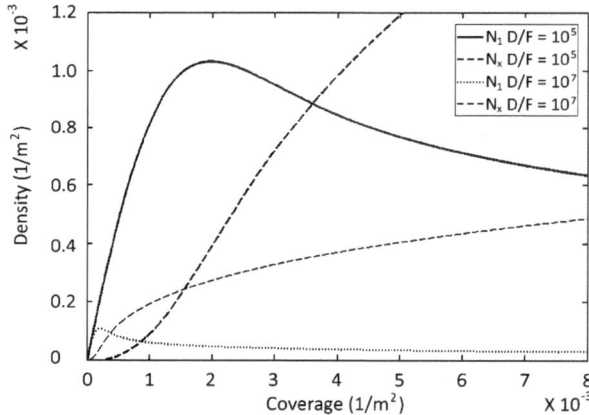

The quantity $f_i(s/S)$ is the scaling function for the island size distribution corresponding when the critical sized island is equal to i.

The variation of the scaled island size distribution with the scaled island size is reported by Shi et al. (2005). A comparison of the data generated in the present work against Shi et al. (2005) is shown in Fig. 2.8. A close match between the two is obtained. The variation of monomer density and saturation island density with coverage in Fig. 2.9 also show a good match.

2.3.4 Parametric Study with Atomistic Model

After validation, a parametric study has been carried out for studying the variation in monomer density, saturation island density, and density of stable clusters with respect to parameters D, F, and τ_{ads}. Simulation is conducted for the limiting case of

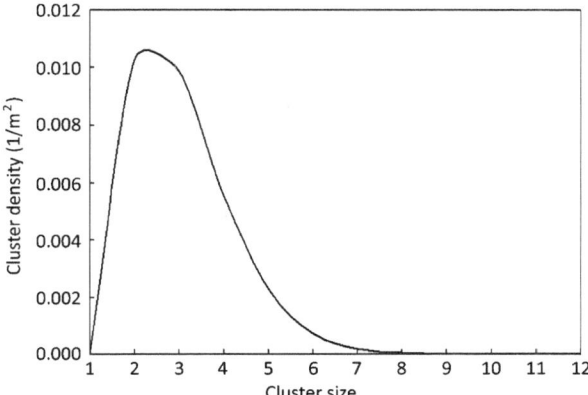

Fig. 2.10 Variation of the number density of clusters with their size

zero flux deposition rate ($F = 0$). The results, plotted in Fig. 2.10, show that the initial spike in the number density distribution vanishes when the deposition rate is zero. The number density distribution of the condensing clusters on the substrate as a function of the model parameters D, F, and τ_{ads} is shown in Fig. 2.11a, b. The first peak at the origin of the coordinate system corresponds to single adatoms originating from the impingement of the vapor flux. The second peak indicates the most probable cluster size of the condensate. The tail of the distribution shows that sizes beyond a certain value do not appear on the substrate. The size distribution determined from (2.1–2.3) is purely from microscopic considerations and does not include macroscopic influences such as surface tension and gravity. Hence, the largest cluster, corresponding to the smallest number density in Fig. 2.11a, b can be interpreted as the smallest drop that would appear on a macroscopic viewpoint. Beyond this size, factors such as gravity, surface tension, and coalescence would be operative in determining the increase in drop diameter.

The preceding expectation has been examined with reference to the thermodynamic estimate (2.24) as follows. At atmospheric pressure and a surface maintained at 80 °C, one can calculate $r_{min} = 9.617 \times 10^{-10}$ m for water. The number of molecules in the drop can be found from the relationship,

$$n_d = \frac{N_A \cdot \pi \cdot r^3}{3 \cdot \bar{M} \cdot v_l} (2 - 3 \cos \theta + \cos^3 \theta) \tag{2.9}$$

Using properties of water, namely, molecular weight $\bar{M}$ of 18 g/mol, N_A the Avogadro number and $\theta = 90°$, we get,

$$n_d = \left(2\pi \cdot r^3 \cdot N_A\right)/\left(3 \cdot \bar{M} \cdot v_l\right) \tag{2.10}$$

The volume referred in (2.9) and (2.10) is that of the spherical cap of a droplet whose radius is r_{min} (see 2.24) and contact angle is θ. The number of molecules

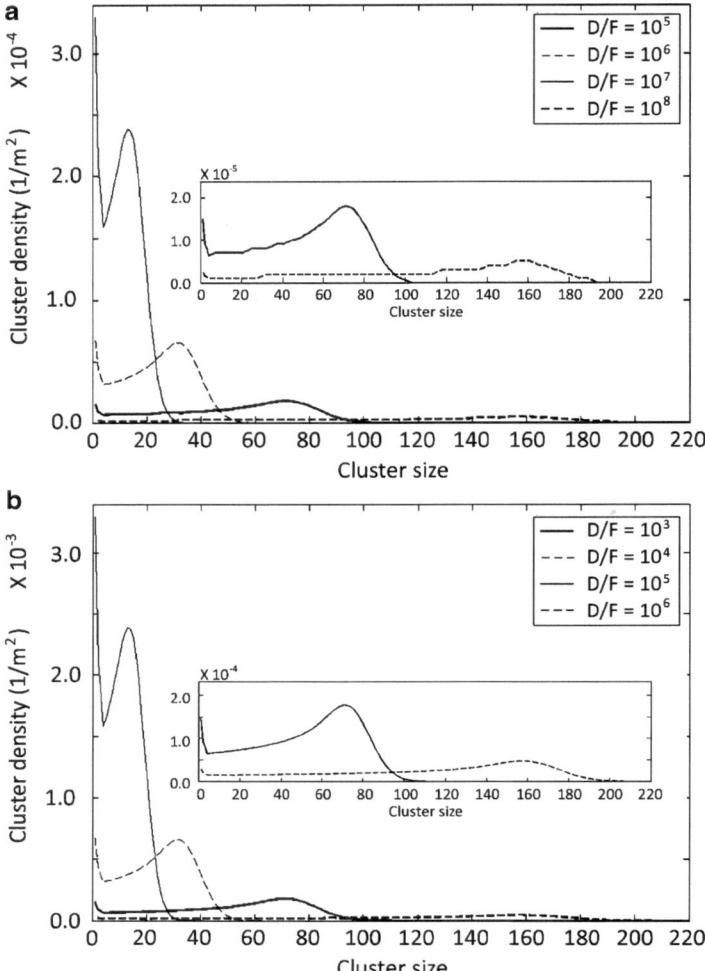

Fig. 2.11 (a) Variation of cluster density with cluster size at $F = 0.005$ s^{-1}. The cluster size where the number density becomes zero yields the maximum cluster size. *Inset* shows the details of the island density profiles for $D/F = 10^7$ and 10^8. (b) Variation of the cluster density with cluster size at $F = 0.05$ s^{-1}. *Inset* shows the details of the island density profiles for $D/F = 10^5$ and 10^6

corresponding to the minimum radius of 9.617×10^{-10} m can now be estimated as $n_d = 60$. In the cluster model, the following results were obtained:

$D = 5{,}000$ and $F = 0.005$, $n_d = 53$
$D = 500$ and $F = 0.05$, $n_d = 58$
$D = 50$ and $F = 0.5$, $n_d = 62$

The number of molecules thus calculated in the smallest drop corresponds quite well to the data of Fig. 2.8.

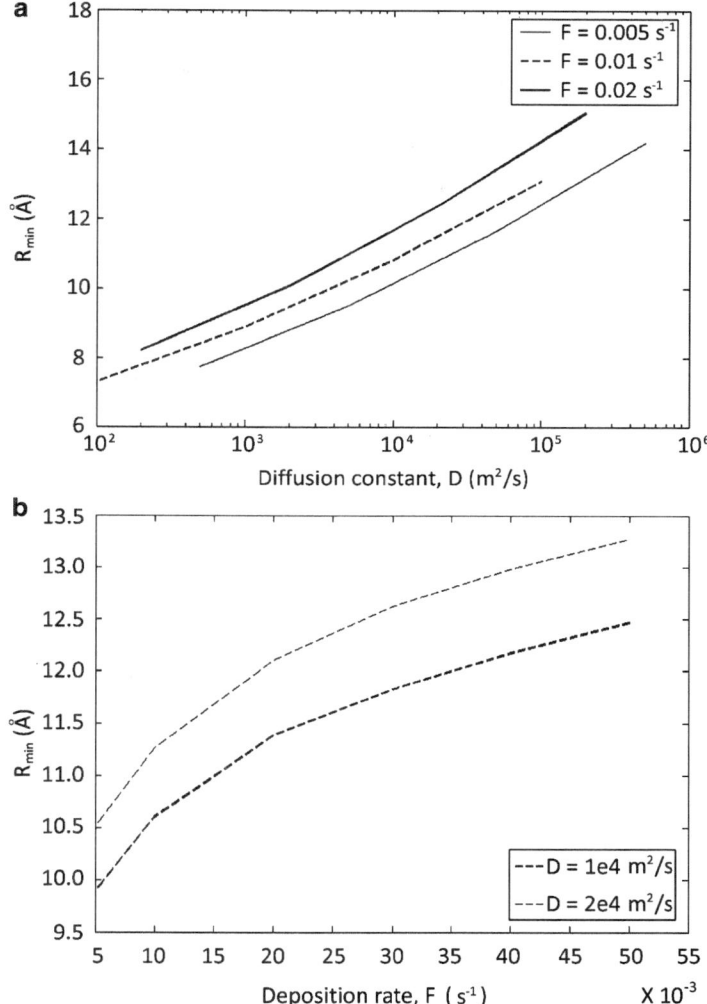

Fig. 2.12 (a) Variation of the minimum drop radius with diffusion constant D at three different deposition rate F. (b) Variation of the minimum drop radius with deposition rate F at two different values of diffusion constant D

The sensitivity of the drop size to the diffusion parameter D and the impinging flux F are shown in Fig. 2.12a, b. The minimum drop size is seen to increase with D as well as F, though the change is not substantial. For an increase of 4 orders of magnitude in the diffusion coefficient, the minimum drop radius increases by a factor of about 2. For an increase of 1 order of magnitude in the vapor flux, the minimum drop radius increases by about 30 %.

These changes are related to the slight broadening of the cluster density and hence there is an increase in the size of the largest possible cluster. A higher mass

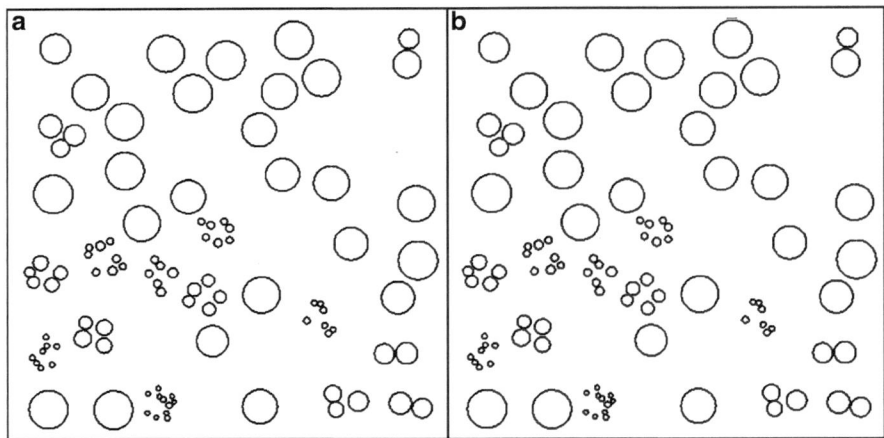

Fig. 2.13 Drop size distribution on a surface for (**a**) $r_{min} = 10$ Å and $r_{max} = 5$ mm, (**b**) $r_{min} = 100$ Å and $r_{max} = 5$ mm during condensation of water vapor

flux increases the number density of adatoms over the substrate and consequently diminishes the extent of diffusion away from the clusters. A higher diffusion constant encourages the association process of monomers and permits the clusters of larger sizes to form. Both the factors lead to an increase in the number of molecules in the largest cluster and hence, in the minimum drop radius.

Apart from the material properties of the condensing medium, the diffusion coefficient is a function of the surface properties and temperature of the substrate. The vapor flux is a process parameter and can be independently controlled. The cluster model given by (2.1–2.3) predicts that by varying D, in effect varying the surface properties, the minimum drop radius is altered. One method available for altering the surface characteristics is physical texturing. As discussed by Chen et al. (1996)

$$D \propto 1/\bar{\eta} \qquad (2.11)$$

The symbol $\bar{\eta}$ represents friction coefficient of the surface. When $\bar{\eta}$ is small,

$$D \propto 1/\bar{\eta}^{0.5} \qquad (2.12)$$

While texturing decreases the friction coefficient, the diffusion coefficient increases, with a corresponding increase in the minimum drop diameter (Chen et al. 1996, 1999). The increase is, however, marginal, as shown in Fig. 2.11a. For chemical texturing of a surface, first principles calculations can be used to predict the diffusion constant (Bloch et al. 1993; Ratsch et al. 1997).

The sensitivity of the drop size distribution on the macroscale to the initial minimum drop radius is examined in Fig. 2.13. Here, the question of special interest is whether the drop size distribution can be influenced by controlling the minimum drop radius. To answer this question, two different r_{min} values were started with, and droplet growth simulation was carried out till drops were large enough for fall-off. The two distribution patterns which emerge are practically identical, suggesting that the macroscale drop distribution is determined by coalescence dynamics, rather than the minimum drop radius.

2.4 Macroscopic Modeling of Dropwise Condensation

Dropwise condensation at the macroscale is a consequence of the time dependent subprocesses associated with the formation of drops at nucleation site, growth by direct condensation and coalescence, sliding motion, fall-off, and then by renucleation on or underneath the substrate. It is a complex intricately linked phenomenon. A mathematical model of these subprocesses is required to describe the entire dropwise condensation process (Sikarwar et al. 2013a, b).

Atomistic modeling of drop formation reveals that the maximum stable cluster obtained by atomistic nucleation process is equal to the size of the minimum stable radius obtained from thermodynamics consideration. Simulations show that condensation patterns at longer timescales are not sensitive to the atomic level processes that fix the minimum drop radius. Therefore, atomic level modeling of condensation is dispensed with and drops formed at the initial nucleation sites are directly assigned the minimum possible stable radius from thermodynamic considerations. The expression for the minimum radius is derived first.

2.4.1 Determination of Minimum Droplet Radius

Consider a system, shown in Fig. 2.14a, containing a liquid droplet of radius r_{min} in equilibrium with supersaturated vapor held at constant temperature (T_{w}) and pressure (p_{v}). Vapor is supercooled or in a supersaturated state without a phase transformation since condensation transfers heat to the adjacent wall. The vapor temperature is equal to the condensing wall temperature and the saturation temperature (T_{s}) corresponding to vapor pressure (p_{v}) is higher than T_{w}. The liquid and vapor state for a liquid droplet in equilibrium with surrounding vapor in a phase diagram. "EF" is supercooled and "FG" is super heated state of vapor at a given wall temperature, Fig. 2.14b. Similarly, "AB" is supercooled and "BC" is the superheated state of liquid. At equilibrium, temperature and chemical potential in the vapor and droplet must be the equal. Therefore,

$$\varphi_{\mathrm{ve}} = \varphi_{\mathrm{le}} \qquad (2.13)$$

The pressure in the two phases (liquid and vapor) are related through the Young-Laplace equation

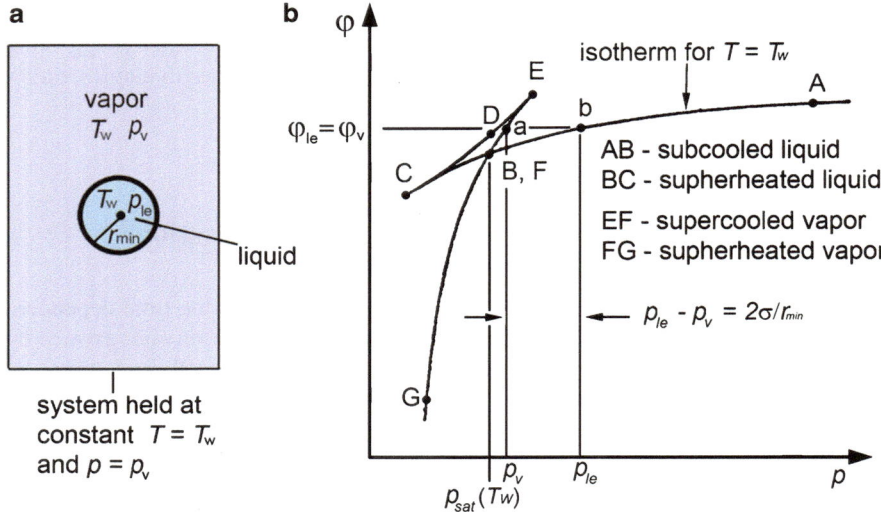

Fig. 2.14 (a) System considered in the analysis of the smallest possible stable droplet. (b) Liquid is in equilibrium with the surrounding vapor

$$p_{\mathrm{le}} = p_{\mathrm{v}} + \frac{2 \cdot \sigma}{r_{\min}} \qquad (2.14)$$

The chemical potential of vapor and liquid phases at equilibrium is evaluated by using the integrated form of the Gibbs-Duhem equation for a constant temperature process. Therefore

$$\varphi - \varphi_{\mathrm{sat}} = \int_{P_{\mathrm{sat}}}^{p} v \, dp \qquad (2.15)$$

We evaluate the integral on the right side (2.15) using the ideal gas law ($v = RT_{\mathrm{w}}/p$) for vapor. Therefore, the chemical potential of the vapor phase is

$$\varphi_{\mathrm{ve}} = \varphi_{\mathrm{sat,v}} + RT_{\mathrm{w}} \ln\left[\frac{p_v}{p_{\mathrm{sat}}(T_{\mathrm{w}})}\right] \qquad (2.16)$$

For the liquid phase inside the droplet, the chemical potential can again be evaluated using (2.15). The liquid is taken to be incompressible, with v equal to the value for saturated liquid at T_{w}. With this assumption, the chemical potential of liquid phase is

$$\varphi_{\mathrm{le}} = \varphi_{\mathrm{sat,l}} + v_{\mathrm{l}} \ln[p_{\mathrm{le}} - p_{\mathrm{sat}}(T_{\mathrm{w}})] \qquad (2.17)$$

Equating the values of φ_{ve} and φ_{le} given by (2.16) and (2.17) to satisfy (2.13), and using the fact $\varphi_{sat,v} = \varphi_{sat,l}$, one obtains

$$p_v = p_{sat}(T_w)\exp\left\{\frac{v_l[p_{le} - p_{sat}(T_w)]}{RT_w}\right\} \tag{2.18}$$

As seen in Fig. 2.14b, if the vapor state point is on the metastable supercooled vapor curve at point a, the liquid state corresponding to equal φ must lie on the subcooled liquid line at point b. For the liquid droplet with finite radius, equilibrium can be achieved only if the liquid is subcooled and the vapor is supersaturated relative to its normal saturation state for a flat interface. Equation (2.18) indicates that if p_v is greater than $p_{sat}(T_w)$, then p_{le} must also be greater than $p_{sat}(T_w)$, consistent with the state points in Fig. 2.14b. Substituting (2.14) to eliminate p_{le}, (2.18) becomes

$$p_v = p_{sat}(T_w)\exp\left\{\frac{v_l[p_v - p_{sat}(T_w) + 2\cdot\sigma/r_{min}]}{RT_w}\right\} \tag{2.19}$$

In most instances, the steep slope of the subcooled vapor line in Fig. 2.14b results in values of p_v that are much closer to $p_{sat}(T_w)$ than p_{le}. Therefore, $p_v - p_{sat}(T_w) \ll 2\sigma/r_{min}$ and (2.19) is well approximated as

$$r_{min} = \frac{2\sigma}{(RT_w/v_l)\ln\left[\frac{p_v}{p_{sat}(T_w)}\right]} \tag{2.20}$$

The Clapeyron equation is combined with the ideal gas law of vapor to obtain

$$\frac{dp}{dT} = \frac{ph_{lv}}{RT_w^2} \tag{2.21}$$

Integrating (2.21) between the p_v and p_{sat} and rearranging

$$\int_{P_{sat}(T_w)}^{p}\frac{dp}{p} = \frac{h_{lv}}{RT_w^2}\int_{T_w}^{T_{sat}}dT \tag{2.22}$$

$$\ln\left[\frac{p_v}{p_{sat}(T_w)}\right] = \frac{h_{lv}}{RT_w^2}(T_{sat} - T_w) \tag{2.23}$$

Substituting (2.23) in (2.20) yields

$$r_{min} = \frac{2\sigma v_l T_w}{h_{lv}(T_{sat} - T_w)}$$ (2.24)

This is the smallest droplet possible corresponding the equilibrium conditions for a specified subcooling of $(T_{sat} - T_w)$.

2.4.2 Nucleation Site Density

The initial, thermodynamically determined drops have a diameter of the order of a few nanometers for fluids encountered in heat transfer applications. Therefore, from an engineering standpoint, it is difficult to experimentally capture the initial nucleation phenomenon on a surface freshly exposed to vapor. Nucleation site density is itself influenced by the thermophysical properties of the condensing fluid, physicochemical properties of the substrate, degree of subcooling, and the substrate morphology. Thus, it is also difficult to determine the nucleation site density on a substrate, either from theory or from experiments. Leach et al. (2006) reported initial site densities close to 10^6 cm^{-2} for temperature differences in the range of 50–100 °C. For condensation of water at 30 °C, the authors suggested that the initial nucleation site density is in the range of 10^4 to 10^5 cm^{-2}, and gradually increases to 10^6 cm^{-2} before the first coalescence. Earlier, a theoretical expression for nucleation site density (in units of cm^{-2}) over an untreated surface was given by Rose (1976) as

$$N = (0.0037)/r_{min}^2$$ (2.25)

Here, N is the number of sites on the substrate per unit area where the initial drops, identifiable as liquid, are formed. Zhao and Beysens (1995) observed no significant connection between the initial nucleation site density and the wettability of the condensing fluid. Rose (2002) indicated that the parameter N in the range of 10^5 to 10^6 is close to the experimental data of dropwise condensation underneath a chemically textured substrate. Mu et al. (2008) found that the nucleation density varies with surface topography, the rougher substrate resulting in a higher nucleation density. Based on the work of Rose (1976, 2002) and Mu et al. (2008), one can conclude that the nucleation density is influenced not only by the degree of surface topography but also by the extent of subcooling. Nucleation density might be influenced by these two factors, i.e., changes in surface energy induced by a chemical species (chemical texturing) and varying roughness morphology of the substrate (physical texturing). The modified expression for the nucleation density of a textured substrate can be expressed as

$$N_f = f \cdot N$$ (2.26)

Here, N_f is the nucleation site density of the textured substrate, f is the degree of roughness and N is initial nucleation density of a smooth surface, as calculated by (2.25), Rose (1976). For a general textured substrate—physical or chemical—factor f needs to be established and is a topic of research.

2.4.3 Nucleation Site Distribution

Heterogeneous nucleation is an important process for phase transitions, including the initial droplet formation during the dropwise condensation process. The initial droplets form only at the natural nucleation sites on the condenser surfaces, and the number of nucleation sites significantly influence the dropwise condensation heat transfer rate. On the other hand, the number of nucleation sites is directly related to the surface properties—its texture, topography, and surface energy distribution. Thus, it is important to study the relationship between surface topography and number of active nucleation sites available at any given instance.

Many researchers have investigated the problem of nucleation site density of dropwise condensation. Glicksman and Hunt (1972) numerically simulated nucleation, growth, coalescence, and renucleation of drops ranging in size from the smallest nucleating drops to the departing drops. Their simulated results agreed well with the data of Krischer and Grigull (1971). Wu and Maa (1976) used the population balance model to derive the size distributions for small pre-coalescence drops. Their calculations showed that the nucleation site density was around $N = 2 \times 10^7$ cm^{-2}. Graham (1969) and Graham and Griffith (1973) studied the nucleation site density with optical microscope photographs. Their results indicated that the site density was 2×10^8 cm^{-2}. Tanasawa et al. (1974) investigated the nucleation site density with electron microscope photographs. The density exceeded 10^{10} cm^{-2} in these measurements. Rose (1976) computed nucleation densities as 5.9×10^9 and 2.9×10^{11} cm^{-2} respectively for minimum nucleation radii r_{min} being 0.07 and 0.01 μm. Leach et al. (2006) reported initial drop densities close to 10^6 cm^{-2} for temperature differences in the range of 50–100 °C. For condensation of water at 30 °C, the initial nucleation site density is in the range of 10^4 to 10^5 cm^{-2} and gradually increases to 10^6 cm^{-2} before the first coalescence. These numbers from various researchers show that the nucleation site density may not be determined by r_{min} alone.

The differences in the nucleation sites densities may also result from the methods used to evaluate the parameters from the images. There are quite few studies related to the nucleation step of dropwise condensation getting influenced by surface characteristics. McCormick and Westwater (1965) applied an optical microscope and showed that the drops nucleated not only at natural cavities on the condenser surface but also at those produced by needles and by erosion and scratches on the surface. Therefore, surface properties of the material affect nucleation site density. Fractal dimension can be used to describe the irregularity and complexity of a rough surface. Yang et al. (1998) and Wu et al. (2001) showed that the droplet distribution had self-similarity, an important feature of fractal behavior. However, the authors studied only the fractal character of droplet pattern without considering the fractal behavior of the condensation surface. In the present work, nucleation site density has been parametrically varied to gage its sensitivity on the resulting heat transfer rate.

From the view point of the current model, the nucleation sites are randomly distributed over the substrate area by using a random seed generator function in C++. The function returns a matrix containing pseudo random numbers with a uniform probability density function in the range [0, 1]. The distribution of sites over the area proceeds column-wise till all the sites are occupied. Once this distribution is carried out, it remains fixed for a given simulation.

Parameters, including the average contact angle, contact angle hysteresis, and the nucleation site density of chemically textured surfaces can be quite different from the physically textured counterparts. These parameters are an input to the condensation model reported in the present study.

Physically textured surfaces are unique in many different ways for the following reasons. For a single drop of liquid sitting over a physically patterned surface, multiple droplet configurations are possible, making the determination of the apparent contact angle a challenge. For example, the static drop could exhibit wetting transitions between Cassie state or the Wenzel state (Berthier 2008; Miljkovic et al. 2012); such configurations of droplets over physically textured surfaces can be seen, for example, in Berthier (2008), Ma et al. (2012) and Rykaczewski 2012). However, for a continuous cyclic process of dropwise condensation on randomized hydrophobic textured surfaces encountering an ensemble of drops of various sizes, it may be argued that the bulk behavior of physically and chemically textured surfaces could be comparable, except for some differences in apparent dynamic contact angles and the mobility of the three-phase contact line. These differences should be small when the drop size is large in comparison to the characteristic scale of the surface roughness. As the drop size at criticality is of the order of a few millimeters, the drop is much larger than the surface features and therefore, the proposed condensation model is expected to be uniformly valid for physically as well as chemically textured surfaces.

2.4.4 Growth by Direct Condensation

Geometric parameters of a drop located underneath textured surfaces of various orientations are estimated. A drop with radius r underneath a horizontal substrate is considered as a part of sphere of contact angle θ, Fig. 2.15a. The contact angle θ on the coated surface is assumed to be constant regardless of the drop size r and the vapor and surface temperature. Therefore, the average contact angle is $\theta_{avg} = \theta$.

For a horizontal substrate with wettability gradient and inclined substrate without wettability gradient, the drop gets deformed and is not a part of a spherical frustum, Fig. 2.15b, c. In these cases, the geometric parameters of a deformed drop

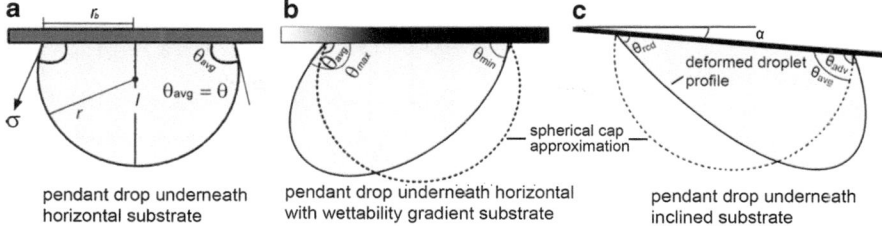

Fig. 2.15 Drop shape underneath (**a**) horizontal substrate, (**b**) horizontal substrate with unidirectional wettability gradient, and (**c**) inclined substrate. For an inclined substrate, drops deform according to advancing angle (θ_{adv}) and receding angle (θ_{rcd}) of the liquid–substrate combination

are calculated using the spherical cap approximation. It is assumed that volume and areas of the deformed drop are equivalent to the part of sphere of contact angle θ_{avg}, as shown in Fig. 2.15b, c. There is some conflict in the calculation of volume of deformed drop and its experimental validation for sessile drops on an inclined surface.

Dussan (1985) and Elsherbini and Jacobi 2004a, b suggested that approximating the drop shape as spherical cap can lead to 10–25 % errors in volume. Based on experimental evidence, others (Extrand and Kumagai 1995; Dimitrakopoulos and Higdon 1999) believe that such approximation is quite valid for a small drop. As pendant drops tend to be small, the spherical cap approximation is used in the present work.

The drops deform according to the applicable value of wettability gradient underneath the horizontal substrate. Therefore, the θ_{avg} of a given ith drop is

$$(\theta_{avg})_i = (1/2)\left[(\theta_{max})_i + (\theta_{min})_i\right] \qquad (2.27)$$

Here, the $(\theta_{max})_i$ and $(\theta_{min})_i$ are contact angles at the two sides of the drop, Fig. 2.15b. For a horizontal substrate with wettability gradient, $(\theta_{max})_i$ and $(\theta_{min})_i$ vary according to the drop position.

Hence, the average contact (θ_{avg}) angle is given as

$$\theta_{avg} = (1/2)(\theta_{rcd} + \theta_{adv}) \qquad (2.28)$$

Here, θ_{rcd} and θ_{adv} are assumed to be constant regardless of the position of drop on the substrate. The drop volume V, area of liquid–vapor interface A_{lv}, maximum drop height from the free surface to wall l, base radius r_b, and area of the solid–liquid interface A_{sl}, are given by the following expressions

$$V = \left(\pi r^3/3\right)\left(2 - 3\cos\theta_{avg} + \cos^3\theta_{avg}\right) \qquad (2.29)$$

$$A_{lv} = 2\pi r^2(1 - \cos\theta_{avg}) \qquad (2.30)$$

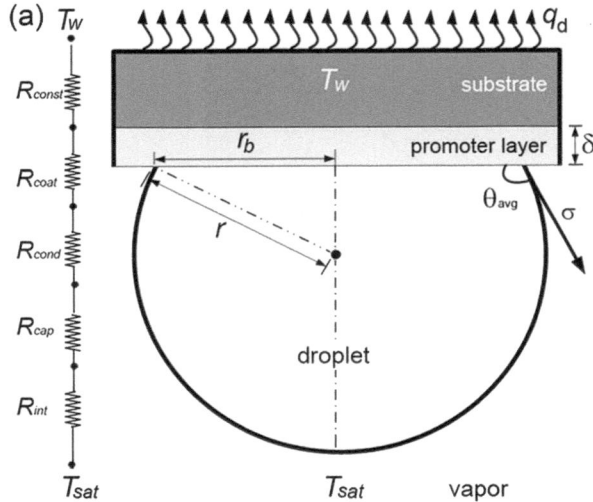

Fig. 2.16 Schematic diagram of a pendant drop with thermal resistances in the droplet growth equation. The promoter layer has a thickness δ, T_{sat} is the vapor saturation temperature, and T_w, the wall temperature

$$A_{sl} = \pi r^2 (1 - \cos^2 \theta_{avg}) \qquad (2.31)$$

$$l = r(1 - \cos \theta_{avg}) \qquad (2.32)$$

$$r_b = r(1 - \cos^2 \theta_{avg}) \qquad (2.33)$$

2.4.4.1 Temperature Drop Due to Various Thermal Resistances

In the proposed model, condensation occurs only over the free surface of the drops. The latent heat release at the free surface is transferred through the volume of liquid to the cold substrate. The substrate area between drops is inactive with respect to heat transfer.

A drop of contact angle θ_{avg} with radius r underneath a textured substrate, which is coated with a promoter layer of thickness δ, shows various thermal resistances in the path of heat transfer, Fig. 2.16. The rate of condensation on the free surface depends on its ability to transfer latent heat released to the cooler substrate. The following thermal resistances are considered in the model:

1. Interfacial resistance (R_{int}) associated with liquid–vapor interface.
2. Capillary resistance (R_{cap}) indicating a loss of driving temperature potential due to droplet interface curvature.
3. Conduction resistance (R_{cond}) associated with the conduction of heat through the droplet.

4. Drop promoter layer resistance (R_{coat}) associated with the thickness of the promoter layer.
5. Constriction resistance (R_{const}) associated with the thermal conductivity of the substrate and nonuniform temperature distribution on the condensing wall due to variable size of drops.

The total temperature difference between the vapor and the substrate ($T_{sat} - T_w$) is the sum of temperature drops due to the individual resistances.

$$(T_{sat} - T_w) = \Delta T_{cond} + \Delta T_{int} + \Delta T_{cap} + \Delta T_{coat} + \Delta T_{const} \qquad (2.34)$$

The component temperature drops are determined as follows:

(a) *Temperature drop due to interfacial resistance*: During dropwise condensation, various researchers have observed that there is transport of molecules crossing the liquid–vapor interface in both directions (Carey 2008). When condensation occurs, the flux of vapor molecules joining the liquid must exceed the flux of liquid molecules escaping into the vapor phase. If the temperature of the interface is equal to the saturation temperature, no net condensation must take place. For net condensation to occur, there should be a finite difference between the saturation temperature of vapor, T_s and temperature of liquid–vapor interface, T_{int}. The temperature difference ($\Delta T_{int} = (T_s - T_{int})$) due to film resistance at the vapor–liquid interface is obtained as

$$\Delta T_{int} = \frac{q_d}{A_{lv} h_{int}} \qquad (2.35)$$

Therefore,

$$\Delta T_{int} = \frac{q_d}{2\pi r^2 (1 - \cos \theta_{avg}) h_{int}} \qquad (2.36)$$

Here, h_{int} is the interfacial heat transfer coefficient, which is usually large and strongly depends on the vapor pressure. It is thus possible to transfer thermal energy at high heat flux levels with relatively low driving temperature difference in a phase change process.

To determine h_{int}, we consider the liquid–vapor interface at the molecular level as shown in Fig. 2.17. The motion of vapor molecules in the vicinity of the interface plays a central role in heat flux limitation during the condensation process. According to the kinetic theory of gases, the statistical behavior of vapor at a certain temperature is described by the Maxwellian velocity distribution

$$\frac{dn_{uvw}}{n} = \left(\frac{m}{2\pi k_b T} \right)^{3/2} e^{-m(u^2 + v^2 + w^2)/2k_b T} \, du \, dv \, dw \qquad (2.37)$$

Fig. 2.17 Liquid–vapor
interface and mass fluxes at
the liquid–vapor interface

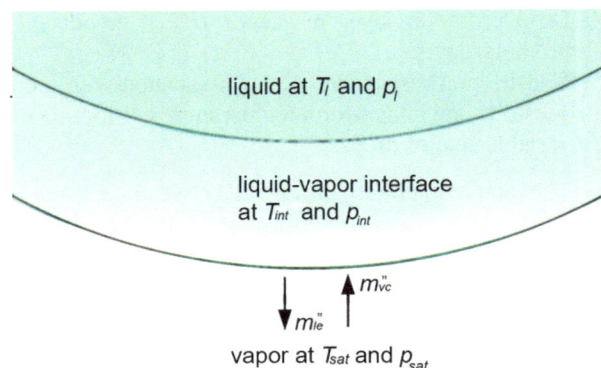

Fig. 2.17 Liquid–vapor interface and mass fluxes at the liquid–vapor interface

If the velocity of vapor molecules obeys the Maxwell distribution, the total rate at which molecules passes through the surface A_{lv} per unit mass and per unit area is

$$j_n = \left(\frac{\bar{M}}{2\pi\bar{R}}\right)^{1/2} \frac{p}{mT^{1/2}} \qquad (2.38)$$

This result of kinetic theory of gases can be used to interpret the motion of vapor molecules near a liquid–vapor interface.

The mass flux of vapor molecules from the vapor phase that impinge on the surface is

$$m''_{\text{vc}} = m \cdot \hat{\sigma} \cdot \Gamma \cdot j_n \qquad (2.39)$$

$$m''_{\text{vc}} = \hat{\sigma} \cdot \Gamma \left(\frac{\bar{M}}{2\pi\bar{R}}\right)^{1/2} \cdot \left(\frac{p_{\text{sat}}}{T_{\text{sat}}^{1/2}}\right) \qquad (2.40)$$

The term Γ corresponds to the fact that the vapor as a whole progresses towards the substrate as long as net condensation takes place. This progress velocity should be superimposed on the Maxwell velocity distribution. The net mass flux of vapor molecules in the direction opposite to the substrate is

$$m''_{\text{le}} = m\hat{\sigma}j_n \qquad (2.41)$$

$$m''_{\text{le}} = \hat{\sigma}\left(\frac{\bar{M}}{2\pi\bar{R}}\right)^{1/2} \cdot \left(\frac{p_{\text{int}}}{T_{\text{int}}^{1/2}}\right) \qquad (2.42)$$

The net mass flux per unit area (m''_{int}) condensing at the liquid–vapor interface is equal to the difference between m''_{vc} and m''_{le}

$$m''_{int} = \left(\frac{\bar{M}}{2\pi\bar{R}}\right)^{1/2}\left[\left(\hat{\sigma}\cdot\Gamma\frac{p_{sat}}{T_{sat}^{1/2}}\right) - \left(\hat{\sigma}\cdot\frac{p_{int}}{T_{int}^{1/2}}\right)\right] \tag{2.43}$$

where

$$\Gamma = 1 + \left[m''_{int}/\left(p_v\sqrt{2\bar{M}/\pi\bar{R}T_{sat}}\right)\right] \tag{2.44}$$

Combining (2.43) and (2.44), we get,

$$m''_{int} = \left(\frac{2\hat{\sigma}}{2-\hat{\sigma}}\right)\left(\frac{\bar{M}}{2\pi\bar{R}}\right)^{1/2}\left(\frac{p_{sat}}{T_{sat}^{1/2}} - \frac{p_{int}}{T_{int}^{1/2}}\right) \tag{2.45}$$

Equation (2.45) can be put into the following form if $((T_{sat} - T_{int})/T_{sat}) \ll 1$:

$$m''_{int} = \left(\frac{2\hat{\sigma}}{2-\hat{\sigma}}\right)\left(\frac{\bar{M}}{2\pi\bar{R}T_{sat}}\right)^{1/2}p_{sat}\left(\frac{p_{sat} - p_{int}}{p_{sat}} - \frac{T_{sat} - T_{int}}{2T_{sat}}\right) \tag{2.46}$$

When the two terms in the parenthese on the right-hand side of (2.46) are compared, the first term, $(p_{sat} - p_{int})/p_{sat}$, is usually much larger than the second, $(T_{sat} - T_{int})/2T_{sat}$. Thus, (2.46) is written as

$$m''_{int} = \left(\frac{2\hat{\sigma}}{2-\hat{\sigma}}\right)\left(\frac{\bar{M}}{2\pi\bar{R}T_{sat}}\right)^{1/2}(p_{sat} - p_{int}) \tag{2.47}$$

Further, from the Clausius-Clapeyron relation

$$\frac{p_{sat} - p_{int}}{T_{sat} - T_{int}} = \frac{\rho_{lv}h_{lv}}{T_{sat}} \tag{2.48}$$

Hence, (2.47) becomes

$$m''_{int} = \left(\frac{2\hat{\sigma}}{2-\hat{\sigma}}\right)\left(\frac{\bar{M}}{2\pi\bar{R}T_{sat}}\right)^{1/2}\frac{\rho_{lv}h_{lv}(T_{sat} - T_{int})}{T_{sat}}; \quad \text{therefore,} \tag{2.49}$$

$$h_{int} = \frac{m''_{int}h_{lv}}{(T_{sat} - T_{int})} \tag{2.50}$$

$$h_{int} = \left(\frac{2\hat{\sigma}}{2-\hat{\sigma}}\right)\left(\frac{h_{lv}^2}{T_{sat}v_{lv}}\right)\left(\frac{\bar{M}}{2\pi\bar{R}T_{sat}}\right)^{1/2} \tag{2.51}$$

Here, the accommodation coefficient ($\hat{\sigma}$) defines the fraction of the striking vapor molecules that actually gets condensed on the vapor–liquid interface. The remaining fraction ($1 - \hat{\sigma}$) is the reflection of vapor molecules that strike the interface but do not condense. The accommodation coefficient indirectly measures the interfacial resistance of the liquid–vapor interface to condensation. Higher the accommodation coefficient, lower the interfacial resistance of the liquid–vapor interface of the condensed drop. For liquid ethanol, methanol, alcohol, and water, the reported values of the accommodation coefficient range from 0.02 to 0.04. On the other hand, reported values for benzene and carbon tetrachloride are closer to unity. It has values ranging from 0.37 to 0.61 for condensation of metallic vapor.

(b) *Temperature drop due to capillary resistance*: As discussed earlier, a pressure difference occurs at the liquid–vapor interface. Therefore, interface temperature is below the saturation temperature of vapor. The depression of the equilibrium interface temperature below the normal saturation temperature for the droplet of radius r can be estimated by replacing ($T_{sat} - T_w$) by ΔT_{cap} and r_{min} by the radius r in (2.24). The resulting relation is given as

$$\Delta T_{cap} = \left(\frac{2\sigma}{r}\right)\left(\frac{v_l T_w}{h_{lv}}\right) = (T_{sat} - T_w)\left(\frac{r_{min}}{r}\right) \tag{2.52}$$

(c) *Temperature drop due to conduction resistance*: The drop itself acts as resistance to heat conduction. Accordingly, the conduction resistance through a liquid drop from the wall to liquid–vapor interface is such that the effective temperature drop associated with this resistance is given by

$$\Delta T_{cond} = \frac{q_d(l/2)}{A_{lv} \cdot k} \tag{2.53}$$

Substituting (2.31 and 2.32) into (2.53) yields the following relation for the temperature drop due to conduction,

$$\Delta T_{cond} = \frac{q_d r(1 - \cos \theta_{avg})}{4\pi r^2 k(1 - \cos \theta_{avg})} \tag{2.54}$$

(d) *Temperature drop due to promoter layer*: The temperature drop due to the resistance offered by the coating material on the substrate is given by

$$\Delta T_{coat} = \frac{q_d \delta}{k_{coat} A_{sl}} \tag{2.55}$$

Substituting (2.30) into (2.55) yields the following relation for the temperature drop

$$\Delta T_{coat} = \frac{q_d \delta}{k_{coat} \pi r^2 (1 - \cos^2 \theta)} \tag{2.56}$$

(e) *Temperature drop due to constriction resistance*: It measures the effect of substrate thermal conductivity on dropwise condensation. There has been continuing controversy about whether the thermal conductivity of the condensing surface plays a significant role in determining effective of heat transfer in dropwise condensation. Results of several investigators (Rose 2002; Bansal et al. 2009) have been interpreted as indicating a strong effect of the thermal conductivity of the substrate on dropwise condensation. But others (Rose 1978a, b) show negligible effect of thermal conductivity of the condensing wall. Tsuruta (1993) has indicated an additional thermal resistance due to the nonuniform heat flux distribution over the condensing surface. In the present simulation, the substrate temperature is assumed uniform. Therefore, the temperature drop due to constriction resistance is absent ($\Delta T_{\text{const}} = 0$).

The total temperature drop will balance the total available subcooling and so,

$$\Delta T_{\text{t}} = \Delta T_{\text{cond}} + \Delta T_{\text{int}} + \Delta T_{\text{cap}} + \Delta T_{\text{coat}} = (T_{\text{sat}} - T_{\text{w}}) \tag{2.57}$$

Therefore, the heat transfer rate through a drop of radius r is obtained as

$$q_{\text{d}} = \frac{(T_{\text{sat}} - T_{\text{w}})(1 - (r_{\min}/r))}{\left(\dfrac{1}{2\pi r^2 h_{\text{int}}(1 - \cos \theta_{\text{avg}})} + \dfrac{r(1 - \cos \theta_{\text{avg}})}{4\pi r^2 k(1 - \cos \theta_{\text{avg}})} + \dfrac{\delta}{\pi r^2 k_{\text{coat}}(1 - \cos^2 \theta_{\text{avg}})} \right)} \tag{2.58}$$

The heat transfer rate through a drop of radius r equals the product of the rate of mass condensate at the free surface and latent heat of vaporization, given by

$$q_{\text{d}} = (\rho_{\text{l}} h_{\text{lv}})(dV/dt) \tag{2.59}$$

$$\frac{dV}{dt} = \frac{dV}{dr} \times \frac{dr}{dt} = \pi r^2 (2 - 3 \cos \theta_{\text{avg}} + \cos^3 \theta_{\text{avg}}) \frac{dr}{dt} \tag{2.60}$$

$$q_{\text{d}} = (\pi r^2 \rho_{\text{l}} h_{\text{lv}}) \times (2 - 3 \cos \theta_{\text{avg}} + \cos^3 \theta_{\text{avg}}) \times \left(\frac{dr}{dt} \right) \tag{2.61}$$

From the above set of equations, one can show that the rate of growth of individual drops follows the equation

$$\frac{dr}{dt} = \left(\frac{4(T_{\text{sat}} - T_{\text{w}})}{\rho_{\text{l}} h_{\text{lv}}} \right) \times \left[\frac{\left(1 - \dfrac{r}{r_{\min}}\right)}{\left(\dfrac{2}{h_{\text{int}}}\right) + \dfrac{r(1 - \cos \theta_{\text{avg}})}{k} + \dfrac{4\delta}{k_{\text{coat}}(1 + \cos \theta_{\text{avg}})}} \right]$$
$$\times \left[\frac{(1 - \cos \theta_{\text{avg}})}{(2 - 3 \cos \theta_{\text{avg}} + \cos^3 \theta_{\text{avg}})} \right] \tag{2.62}$$

Equation (2.62) is valid for horizontal and inclined surfaces without wettability gradient as well as the *i*th drop (with average contact angle of θ_{avg}) underneath a horizontal substrate with wettability gradient.

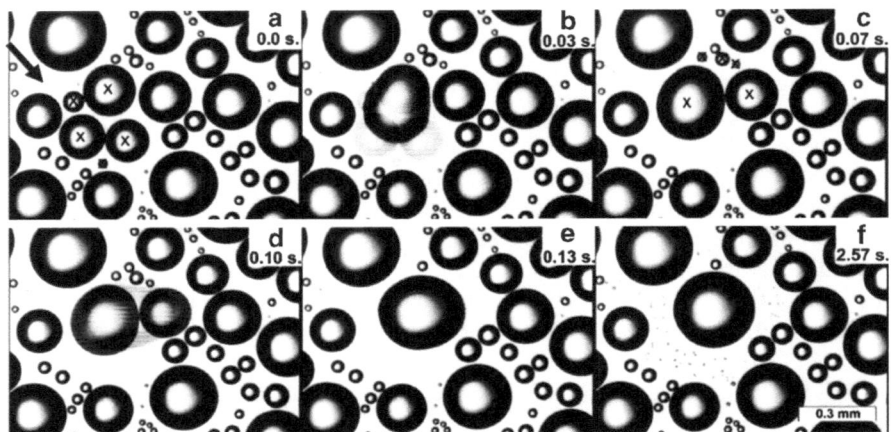

Fig. 2.18 Observing drop coalescence during a dropwise condensation experiment (Leach et al. 2006). (**a**) Drops prior to coalescence; the drop coalescence is marked with ×. The *arrow* identifies a drop-free region where drops will be observed later. (**b**) The same region during coalescence. (**c**) The next images, where drops coalesce are marked with ×. (**d**) The same region during coalescence and (**e**) after coalescence. (**f**) The same region seconds later, after newly nucleated drops become visible following coalescence. Some of these drops appear in the region marked in (**a**), which was drop-free prior to the coalescence events

2.4.5   Growth by Coalescence

In dropwise condensation, two or more drops on or underneath a cold substrate grow large enough to touch one another, coalesce, and form a single larger drop, Fig. 2.18. Leach et al. (2006) studied dropwise condensation of water vapor coming from a hot water reservoir onto a naturally cooled hydrophobic polymer film and a silanized glass slide. The authors observed that the coalesced drop is at the center of mass of the original drops. The smallest detectable droplets were seen to grow and eventually fall-off, after repeated cycles of nucleation to coalescence. The spatio-temporal coalescence scales were also reported. Images acquired before and after coalescence events confirmed that drop coalescence reexposed the substrate area for nucleation of new liquid drops.

Many researchers (Vemuri and Kim 2006; Leipertz 2010; Dietz et al. 2010; Miljkovic et al. 2012) have experimentally recorded the smallest detectable droplets that grow and eventually fall-off after repeated cycles of coalescence. The growth rate of drops depends on their respective size: small drops grow by direct condensation as well as occasional coalescence but large drops grow mainly by coalescence. The growth rate of small drops is related to heat transfer. Smaller drops offer less thermal resistance, thus permitting rapid condensation. Larger drops offer a higher thermal resistance and grow primarily by coalescence. Hence, coalescence plays a primary role in determining the drop size distribution on the macroscale while direct condensation is of secondary importance.

Coalescence also plays a direct role in the frequency of attainment of drop criticality, either for sliding motion or fall-off. Subsequently, nucleation occurs over the reexposed area of the substrate. Nucleation, slide-off or fall-off and droplets coalescence are the fundamental processes that enhance heat transfer coefficient at later stages of growth in dropwise condensation. Since the associated heat transfer rates are high, one can imagine coalescence dynamics as one of the important factors contributing to the enhanced heat transfer during dropwise condensation.

Coalescence-induced instability in the pendant mode is an effective means of passively enhancing heat transfer coefficient during dropwise condensation. Inclined substrates have natural advantage in terms of sweeping of drops from the substrate, thereby, exposing fresh sites for nucleation. As compared to coalescence of sessile droplets, flow instabilities are induced faster in pendant drops, enhancing the associated heat transport characteristics.

Although coalescence of pendant drops underneath an inclined hydrophobic surface is an efficient process in dropwise condensation, discussion on the subject is scarce in the literature. Much of the research available is on the formation of a liquid bridge and the relaxation time coalescence in sessile drops.

Eggers et al. (1999) focused on early-time behavior of the radius of the small bridge between two drops. When two liquid droplets touch each other, a liquid bridge is formed between them. A negative curvature or negative pressure is created at the point of joining. This bridge quickly expands under the influence of interfacial stresses and the resultant fluid motion pulls the two drops together, forming a large drop with a smaller surface area. This motion is viscously dominated in the initial stages. Based on the above concept the authors proposed a scaling law for a variation of the liquid bridge radius with time.

For two drops merging together, Andrieu et al. (2002) experimentally recorded and theoretically described the kinetics of coalescence of two water drops on a plane solid surface. Immediately after coalescence, an ellipsoidal shape results, eventually relaxing into a hemispherical shape, in a few milliseconds. The characteristic relaxation time is proportional to the drop radius R at final equilibrium. This relaxation time is nearly 10^7 times larger than the bulk capillary relaxation time $t_\eta = R\mu/\sigma$, where σ is the vapor–liquid surface tension and μ is the liquid shear viscosity.

Duchemin et al. (2003) studied coalescence of two liquid drops driven by surface tension. The fluid was considered to be ideal and velocity of approach, zero. Using the boundary integral method, the walls of the thin retracting sheet of air between the drops were seen to reconnect in finite time to form a toroidal enclosure. After initial reconnection, retraction starts again, leading to a rapid sequence of enclosures. Averaging over the discrete events, the minimum radius of the liquid bridge connecting the two drops were scaled as r_b proportional to $(t)^{0.5}$.

Using high speed imaging, Wu et al. (2004) studied early-time evolution of the liquid bridge that is formed upon the initial contact of two liquid drops in air. Experimental results confirmed the scaling law that was proposed by Eggers et al. (1999). Further, their experimental study demonstrated that the liquid bridge radius (r_b) follows the scaling law $r_b \propto (t)^{0.5}$ in the inertial region. The pre-factor of the scaling law, $r_b/(t)^{0.5}$, is shown to be proportional to $R^{1/4}$, where R is the inverse of the drop curvature at the point of contact. The dimensionless pre-factor is measured to be in the range of 1.03–1.29, which is lower than 1.62, a pre-factor predicted by the numerical simulation of Duchemin et al. (2003) for inviscid drop coalescence.

Narhe et al. (2004) investigated the dynamics of coalescence of two sessile water drops and compared them with the spreading dynamics of a single drop in the partially wetting regime. The composite drop formed due to coalescence relaxed exponentially towards equilibrium with a typical relaxation time that decreases with contact angle. The relaxation dynamics is larger by 5–6 orders of magnitude than the bulk hydrodynamics which is of the order of a few milliseconds, due to the high dissipation in the contact line vicinity. Narhe et al. (2005) studied the dynamics of drop coalescence in the sessile mode of dropwise condensation of water vapor onto a naturally cooled hydrophobic polymer film and silanized glass slide. The authors reported that coalescence is affected by surface orientation and composition, vapor and surface temperatures, humidity, and vapor flow rate.

Aarts et al. (2005) studied droplet coalescence in a molecular system with variable viscosity and a colloid–polymer mixture with an ultralow surface tension. When either the viscosity is large or the surface tension is small enough, the liquid bridge opening initially proceeds with capillary velocity. Inertial effects are dominant at a Reynolds number of about 1.5 ± 0.5 and the neck then grows as the square root of time. In a second study, decreasing the surface tension by a factor of 10^5 opened the way to a more complete understanding of the hydrodynamics involved.

Thoroddsen et al. (2005) studied pendant as well sessile drop coalescence. The authors used an ultra-high speed video camera to study coalescence, over a range of drop sizes and liquid viscosities. For low viscosity, the outward motion of the liquid contact region is successfully described by a dynamic capillary-inertial model based on the local vertical spacing between two drop surfaces. This model can also be applied to drops of unequal radii. Increasing viscosity slows down coalescence. For the largest viscosity, the neck region initially grows in size at a constant velocity. The authors compared their results with the previously predicted power law, finding slight but significant deviation from the predicted exponents.

Ristenpart et al. (2006) investigated experimentally and theoretically the coalescence dynamics of two spreading drops on a highly wettable substrate. They found that the width of the growing meniscus bridge between the two droplets exhibits power-law behavior, growing at early times as $(t)^{0.5}$. Moreover, the growth rate is highly sensitive to both the radii and heights of the drops at contact, scaling as $h^{3/2}/R_o$. This size dependence differs significantly from the behavior of freely suspended drops, in which the coalescence growth rate depends only weakly on the drop size.

Kapur and Gaskell (2007) experimentally investigated coalescence of a pair of drops on a surface with high quality images from flow visualization revealing the morphology of the process. The drops merge and evolve to a final state with a footprint that is peanut-like in shape, with bulges along the longer sides resulting from the effects of inertia during spreading. The associated dynamics involve a subtle interplay between (a) the motion of the wetting process due to relaxation of the contact angle and (b) a rapid rise in free surface height above the point where coalescence begins due to negative pressure generated by curvature. During the early stages of motion, a traveling wave propagates from the point of initial contact up the side of each drop as liquid is drawn into the neck region, and only when it reaches the apex of each do their heights start decrease. A further feature of the rapid rise in height of the neck region is that the free surface overshoots significantly from its final equilibrium position; it reaches a height greater than that of the starting drops, producing a self-excited oscillation that persists long after the system reaches its final morphological state in relation to its footprint.

Thoroddsen et al. (2007) studied drop coalescence of two different miscible liquids and found that the coalescence speed is governed by the liquid having weaker surface tension. Marangoni waves propagate along the drop with stronger surface tension. Surface profiles and propagation speeds of these waves were reported from experiments with a pendant water drop coalescing with a flat ethanol surface or with a sessile drop of ethanol. In the former, capillary-Marangoni waves along the water drop showed self-similar character in terms of arc length along the original surface.

Liao et al. (2008) performed an experimental investigation on coalescence of two equal-sized water drops on inclined surfaces. The effects of inclination angle and the drop size were studied with respect to the liquid bridge, fore/back contact angle and, and the evolving three-phase contact line.

Sellier and Trelluyer (2009) proposed a power law growth of the bridge between the drops describe the coalescence of sessile drops. The exponent of the power law depends on the driving mechanism for the spreading of each drop. The authors validated the experiment against numerical simulation.

Boreyko and Chen (2009) linked the coalescence with heat transfer rate in dropwise condensation. The authors experimentally showed the drop shifting on a substrate and releasing interfacial energy during coalescence. Energy released is higher for higher contact angle and is responsible for the drop movement and enhancement of heat transfer during coalescence.

Wang et al. (2010) conducted an experiment to study the behavior of liquid drop coalescence on a surface with gradient in surface energy. The microscopic contour of the gradient energy surface was fabricated on the base of a silicon chip by diffusion controlled silanization of alkyltrichlorosilanes and characterized by an atomic force microscope. The effect on the three-phase contact line and contact angle was obtained. The process of drop coalescence was seen to accelerate the drop speed on the gradient surface.

Sellier et al. (2011) studied coalescence of sessile drops of distinct liquids, arising from Marangoni stresses due to surface tension gradient. The analysis revealed two dimensionless numbers that govern flow characteristics. One is related to the strength of surface tension gradient and the other to the diffusion timescale. Numerical results confirmed the occurrence of the self-propulsion behavior.

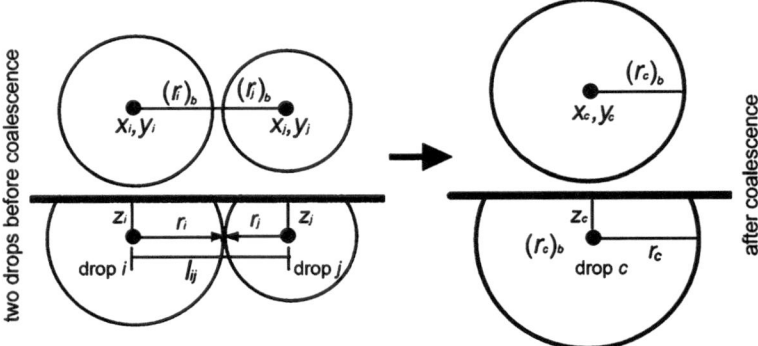

Fig. 2.19 Schematic showing coalescence of two drops and criteria of coalescence

Paulsen et al. (2011) used an electrical method and high-speed imaging to investigate drop coalescence down to 10 ns after the drops touch. Viscosity was varied over two decades. At a sufficiently low approach velocity where deformation is not present, the drops coalesced with an unexpectedly late crossover time between a regime dominated by viscous and one by inertial effects.

Much of the research on the topic covers the formation of a liquid bridge and the relaxation time of sessile drops during coalescence. Coalescence of pendant drops and its role of coalescence in heat transfer enhancement are not readily available.

2.4.5.1 Modeling Growth by Coalescence

A simple model of drop coalescence is adopted in the present work on the basis of experimental observations reported in the literature. Consider two drops of radius r_i and r_j at nucleation sites, i and j, on the substrate, Fig. 2.19.

The distance between the two nucleation sites, i and j, is calculated as

$$l_{ij} = \sqrt{(x_i - x_j)^2 + (y_i - y_j)^2 + (z_i - z_j)^2} \qquad (2.63)$$

The coalescence criterion can be stated as

$$l_{ij} - [r_i + r_j] < 10^{-6} \qquad (2.64)$$

If the coalescence criterion is met, a drop of equivalent volume on the mass averaged center of the original coalescing droplets is introduced. The time for coalescence is taken to be much smaller than the other timescales of the condensation process. Hence, as soon as the two droplets contact each other (or three droplets, or, very rarely, four contact each other simultaneously), they are substituted with an equivalent single drop with equal total volume, located at the weighted center of mass of the individual coalescing drops. Shows two drops i and j in a coalescence process forming drop c, as shown in Fig. 2.19. The volume, position, and radius of drop c are

$$V_c = (V_i + V_j) \tag{2.65}$$

$$x_c = (V_i x_j + V_j x_i)/(V_c) \tag{2.66}$$

$$y_c = \frac{(V_i y_j + V_j y_i)}{V_c} \tag{2.67}$$

$$r_c = \left[\frac{3(V_i + V_j)}{\pi(2 - 3 \cos \theta_{\text{avg}} + \cos^3 \theta_{\text{avg}})} \right]^{1/3} \tag{2.68}$$

The base radius of the drop formed after coalescence is

$$(r_c)_b = r_c \sin \theta_{\text{avg}} \tag{2.69}$$

Equations (2.63–2.69) are valid for horizontal and inclined surfaces with and without wettability gradient. More than two drops (i, j, and k) are coalesced as

$$l_{ij} = \sqrt{(x_i - x_j)^2 + (y_i - y_j)^2 + (z_i - z_j)^2} \tag{2.70}$$

$$l_{ik} = \sqrt{(x_i - x_k)^2 + (y_i - y_k)^2 + (z_i - z_k)^2} \tag{2.71}$$

$$l_{jk} = \sqrt{(x_j - x_k)^2 + (y_j - y_k)^2 + (z_j - z_k)^2} \tag{2.72}$$

The coalescence criterion can be stated as

$$l_{ij} - [r_i + r_j] < 10^{-6}; \qquad l_{ik} - [r_i + r_k] < 10^{-6}; \qquad l_{jk} - [r_j + r_k] < 10^{-6} \tag{2.73}$$

The volume, position, and radius of drop (drop c) formed by coalescing more than two drop is determined as,

$$V_c = (V_i + V_j + V_k) \tag{2.74}$$

$$x_{c,i} = \frac{(V_i x_j + V_j x_i)}{V_i + V_j} \tag{2.75}$$

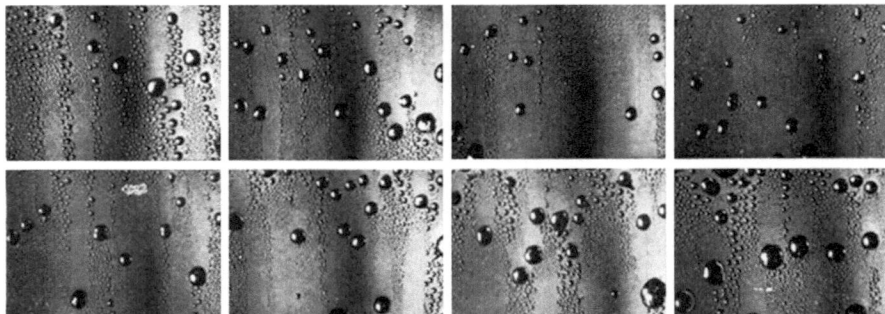

Fig. 2.20 Image of drops siding in dropwise condensation on copper substrate at different inclination with the horizontal (Citakoglu and Rose 1968b)

$$y_{c,i} = \frac{(V_i y_j + V_j y_i)}{V_i + V_j} \tag{2.76}$$

$$x_c = \frac{((V_i + V_j)x_k + V_k x_{c,i})}{V_i + V_j + V_k} \tag{2.77}$$

$$y_c = \frac{((V_i + V_j)y_k + V_k y_{c,i})}{V_i + V_j + V_k} \tag{2.78}$$

$$r_c = \left[\frac{3(V_i + V_j + V_k)}{\pi(2 - 3\cos\theta_{\text{avg}} + \cos^3\theta_{\text{avg}})} \right]^{1/3} \tag{2.79}$$

The main assumption in the approach adopted for coalescence is that its time-scale (in milliseconds) is small in comparison with the cycle time of dropwise condensation (usually in excess of a second). The coalesced drops relax over a longer time period but this process can be neglected because, most often, it would become gravitationally unstable, leading to fall off or slide off from the substrate. Hence, the assumption of instantaneous coalescence is expected to be reasonable in dropwsie condensation.

2.4.6   Drop Instability

When a certain size is reached, several authors (Citakoglu and Rose 1968a, b; Meakin 1992; Leipertz and Fröba 2006, 2008) have shown that the gravitational force on the droplet exceeds the adhesive force between the liquid and condensing substrate, and the droplet begins to move, Fig. 2.20.

Drop motion plays an important role in the enhancement of heat transfer. The sliding drop wipes other droplets off, resulting in 'reexposed' surface area.

New drops are formed again in the reexposed area of the substrate. The diffusional resistance of the liquid contained in these drops forms the primary thermal resistance of the energy released from the free surface to the condensing wall. Enhancement of heat transfer necessitates that these drops be swept away from the substrate as soon as possible so as to reduce the most prominent thermal resistance in the passages of heat, from the vapor to the substrate. Sliding may be achieved either by (a) inclining the substrate, or alternatively, (b) by creating as additional force imbalance at the three-phase contact line.

The later strategy is most suitable for induced motion on horizontal surfaces. Contemporary manufacturing/coating techniques can provide such a wettability gradient by physicochemical action, leading to additional surface forces required for inducing droplet motion.

2.4.6.1 Drop Sliding on an Inclined Substrate

Literature on drop sliding on or underneath a textured inclined surface is limited. Most of the existing work (Brown et al. 1980; Dussan 1985; Briscoe and Galvin 1991b; Elsherbini and Jacobi 2006) have considered the critical state of static sessile drop on an inclined surface and focused on the apparent contact angle hysteresis, drop shape, and drop retention with tiltable surfaces for various combinations of hydrophobic surfaces and liquids.

Though a large volume of work exists on predicting the drop shape under static condition, only a few researchers have reported the sliding behavior of the drop on an inclined surface as well as horizontal wettability gradient substrate.

Kim et al. (2002) reported that a liquid drop which partially wets a solid surface will slide down the plane when it is tilted beyond a critical inclination. Experiments for measuring the steady sliding velocity of different liquid drops were performed on an inclined surface leading to a scaling law to relate velocity with wetting characteristics.

Grand et al. (2005) reported experiments on the shape and motion of millimeter-sized drops sliding down a plane in a situation of partial wetting. An unexpected shape change was seen when the velocity of drop is increased. In theoretical analysis, the viscous force was scaled as $\mu U V^{1/3}$ and the drop sliding velocity was found to be a linear function of the Bond number. Rio et al. (2005) examined the microscopic force balance close to a moving contact line to investigate boundary conditions around viscous drops sliding down an inclined plane.

Gao and McCarthy (2006) postulated two mechanisms for a drop moving down the plane. Drops move by sliding, when the particles near the solid–liquid interface exchange their position with those at the gas–liquid interface, while the bulk of the fluid remains unaffected. On the other hand, there could be rolling motion where the entire fluid mass undergoes a circulatory movement. Sakai et al. (2006) used particle image velocimetry (PIV) to observe the internal fluidity of water droplets during slide on various chemically textured surfaces. On normal hydrophobic surface with contact angle of around 100°, both slipping and rolling controlled velocity during

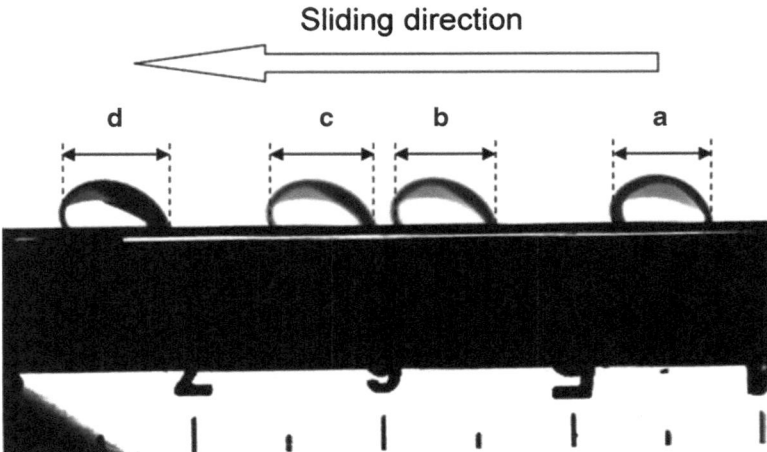

Fig. 2.21 Photographs of the sliding of a 45 mg water droplet on the sample coated with FAS tilted at 35°. Each sliding distance at (**a**), (**b**), (**c**), and (**d**) was 0.000, 0.010, 0.015, and 0.030 m. The droplet length at corresponding distance is 5.28 mm, 5.34 mm, 5.58 mm and 5.78 mm, respectively (Suzuki et al. 2006)

slide. On the superhydrophobic surface, however, with a contact angle of 150°, the droplet fell at high velocity by slipping. Yoshida et al. (2006) did not consider the viscous force in their study of the sliding behavior of water drops on a flat polymer surface. The authors reported that sliding motion changed from constant velocity to one of constant acceleration with an increase in the contact angle. Suzuki et al. (2006) reported a photograph of a sliding 45 mg water droplet on the surface coated with fluoroalkylsilane and tilted at 35°, Fig. 2.21. The authors reported that apparent length of water droplets increases when the sliding velocity increases.

Sakai and Hashimoto (2007) experimentally determined the velocity vector distribution inside a sliding sessile drop using PIV. The authors reported that the velocity gradient near the liquid–solid interface is higher than locations elsewhere inside a drop.

Hao et al. (2010) investigated the internal flow pattern in a water droplet sliding on the superhydrophobic surface by employing PIV and PTV. Both rolling and slipping motion were seen inside the drop during sliding, though rolling occurred only at the edge of the water droplet.

2.4.6.2 Drop Sliding Over Horizontal Surface with Wettability Gradient

The possibility of drop movement resulting from a wettability gradient was noted by Greenspan (1978) and experimentally demonstrated by Chaudhury and Whitesides (1992). Daniel et al. (2001) performed the experiment of condensation

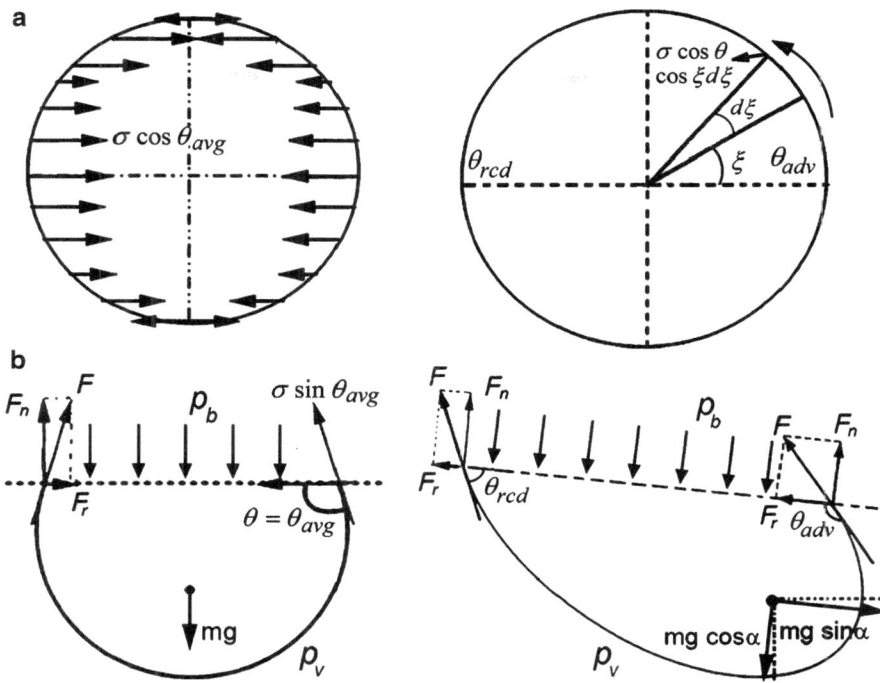

Fig. 2.22 (**a**) Direction of the retention force shown on the footprint and, (**b**) free body diagram of drop underneath a horizontal substrate and an inclined substrate

on a wettability gradient substrate. The authors observed more rapid motion (1.5 m/s) when condensation occurred over a horizontal wettability gradient surface. Drops moved hundreds to thousands of times faster than the speeds of typical Marangoni flows. Moumen et al. (2006) measured the velocity of a drop along a wettability gradient surface. At steady state, the driving force for drop movement comes from the gradient of free energy of adhesion of the drop with the substrate, and balanced by viscous drag generated within the liquid drop.

2.4.6.3   Modeling Drop Instability

During condensation, drops grow first by direct deposition of vapor and then by coalescence. Continuously, the weight of the drop increases and can be a destabilizing influence. Force imbalance at the three-phase contact line leads to instability. For definiteness, a free body diagram of a pendant drop underneath a flat horizontal substrate and an inclined substrate with corresponding forces acting at the three-phase contact line, Fig. 2.22, is considered. For determining the onset of instability, pressure, surface tension, and gravity are taken as dominant forces. The size of drop at slide-off and fall-off are estimated in the following sections.

Liquid pressure within the drop will be in excess of the surrounding vapor pressure. The pressure difference is larger in the smaller drop. As the drop size increases, pressure difference decreases. It is a minimum at the onset of instability, for which the drop diameter has increased to its largest possible size. When the drop is about to slide, pressure acts normal to the surface and does not contribute to the force calculation.

For fall-off, the excess pressure has a component in the vertical direction. Since it is small for the large drops, excess pressure has been neglected in the fall-off instability calculation. Accordingly, the critical drop diameter is expected to be slightly overpredicted.

2.4.6.4 Horizontal Substrate

A pendant drop underneath a horizontal substrate is a part of a sphere of radius r with a contact angle $\theta = \theta_{avg}$. For a horizontal hydrophobic substrate, surface tension and gravity are in competition, Fig. 2.22a. As droplets grow in size, body forces (gravity) eventually surpass the limiting surface force (surface tension) at the three-phase contact line. As discussed above, the contribution of excess pressure in the determination of the critical drop diameter is negligible.

The component of surface tension force normal to the substrate is

$$F_n = 2\pi r_b \sigma \sin \theta_{avg} = 2\pi r \sigma \sin^2 \theta_{avg} \tag{2.80}$$

The weight of the drop is

$$F_{g\perp} = g(\rho_1 - \rho_v)(\pi r^3/3)(2 - 3\cos\theta_{avg} + \cos^3\theta_{avg}) \tag{2.81}$$

Equating (2.80) and (2.81), the maximum radius (the size of droplet fall-off) is obtained as

$$r_{max} = \sqrt{\left(\frac{6\sin^2\theta_{avg}}{(2 - 3\cos\theta_{avg} + \cos^3\theta_{avg})}\right)\left(\frac{\sigma}{(\rho_1 - \rho_v)g}\right)} \tag{2.82}$$

Uniform distribution of the contact angle at the three-phase contact line makes the net retention force (F_r) underneath a horizontal plane zero. The arguments leading to (2.82) do not include adhesion of the liquid with the substrate at the base, since higher order effects on the meso-scale and micro-scales are neglected. Various authors (Amirfazli 2007; Miljkovic et al; 2012; Rykaczewski 2012) showed that such effects are not important on engineering scale calculations.

2.4.6.5 Inclined Substrate

Inclining the substrate causes an imbalance in the forces and results in drop deformation, to achieve necessary static balance. A deformed drop underneath an inclined substrate at incipient sliding is shown in Fig. 2.22b. The leading side contact angle is equal to the advancing angle and trailing side contact angle is the receding angle of the liquid substrate combination. The figure highlights the relevant forces at contact line which is taken to be circular. The force balance on a drop underneath an inclined substrate at incipient sliding is also shown. The component of body force (gravity) parallel to the substrate tries to slide the drop and surface tension provides the retention force for stability. Similarly, body force component normal to the substrate leads to fall-off while the normal component of surface tension provides stability to hold the drop. Hence, the critical size at which slide/fall-off commences depends not only on the thermophysical properties of the liquid but also on physicochemical properties of the substrate.

Under dynamic conditions, the applicability of static force balance is questionable due to the presence of capillary waves, distortion in local equilibrium droplet shapes, droplet pinning, variation in dynamic contact angle due to inertia effects, sudden acceleration, and three-dimensional flow structures inside the droplets. Therefore, there is considerable debate in the literature on the applicability of static conditions on the real-time condensation process (Fang et al. 2008; Annapragada et al. 2012). The bulk composite effect of these real-time dynamic situations and local contact line perturbations is manifested in the form of hysteresis of advancing and receding angles. The static force balance conditions should be representative of the dynamic situation, since absolute contact angles and hysteresis are accounted for. Expressions for the maximum base radius of a drop that will first slide (r_{crit}) or fall-off (r_{max}) underneath an inclined substrate are derived in the following discussion.

1. *Estimation of critical radius of slide-off underneath an inclined substrate*

The critical radius of the droplet at slide-off underneath an inclined substrate is obtained by force balance parallel to the substrate. Accordingly, the retention force arising from contact angle hysteresis, namely, the difference in the advancing angle and receding angle, is equal to the component of weight parallel to the substrate. The component of retention force acting on the drop in the direction of substrate inclination is found by integrating the differential force over the base of the drop as follows:

$$F_{r\parallel} = 2 \int_0^\pi r_b \sigma \cos \theta \cos \xi \, d\xi \qquad (2.83)$$

The normal component of surface tension at the base of the drop is

$$F_{r\perp} = 2 \int_0^\pi r_b \sigma \cos \theta \sin \xi \, d\xi \qquad (2.84)$$

Due to symmetry at the base, $F_{r\perp} = 0$; the resulting retention force (F_r) due to surface tension acts in the direction of substrate inclination ($F_{r\parallel}$).

The contact angle hysteresis, namely, the variation in the advancing to receding contact angle, is taken to vary linearly along the contact line with respect to azimuthal angle. The base of the droplet is taken to be circular as discussed earlier. The variation of contact angle, with respect to azimuthal angle along the drop contact line is formulated as

$$\cos \theta = \cos \theta_{adv} + \left(\frac{\cos \theta_{rcd} - \cos \theta_{adv}}{\pi} \right) \xi \tag{2.85}$$

Substituting the (2.85) into (2.83) and integrating, one obtains

$$F_r = -(4/\pi)\sigma r_b(\cos \theta_{rcd} - \cos \theta_{adv}) \tag{2.86}$$

The ($-$) sign indicates the direction of force is opposite to the direction of inclination. The drop volume (V), area of liquid–vapor interface A_{lv}, and area of solid–liquid interface A_{sl} of deformed drop underneath an inclined substrate are calculated using the spherical cap approximation. Accordingly, the volume of the deformed drop is

$$V = \frac{\pi r_b^3(2 - 3 \cos \theta_{avg} + \cos^3 \theta_{avg})}{3 \sin^3 \theta_{avg}} \tag{2.87}$$

The force component due to gravity that is parallel to the substrate is

$$F_{g\parallel} = \frac{\pi r_b^3(2 - 3 \cos \theta_{avg} + \cos^3 \theta_{avg})}{3 \sin^3 \theta_{avg}} (\rho_1 - \rho_v)g \sin \alpha \tag{2.88}$$

Here, r_b is the base radius of drop and is related to the drop radius as

$$r_b = r \sin \theta_{avg} \tag{2.89}$$

A balance of forces acting direction parallel to substrate inclination yields

$$F_{g\parallel} + F_{r\parallel} = 0 \tag{2.90}$$

Hence, the critical radius of the droplet at slide-off on the inclined substrate is calculated as

$$r_{crit} = \sqrt{\left(\frac{1.25 \sin \theta_{avg}}{(2 - 3 \cos \theta_{avg} + \cos^3 \theta)} \right)(\cos \theta_{rcd} - \cos \theta_{adv})\left(\frac{\sigma}{(\rho_1 - \rho_v)g \sin \alpha} \right)} \tag{2.91}$$

For $r_b > r_{crit}$, the drop becomes unstable and slides over the surface.

2. *Estimation of critical radius of fall-off underneath an inclined substrate*

Surface tension perpendicular to the inclined substrate is calculated as

$$F_n = 2 \int_0^\pi r_b \sigma \, \sin \theta \, d\xi \qquad (2.92)$$

The variation of contact angle, with respect to azimuthal angle (ξ) along the contact line is formulated as

$$\sin \theta = \sin \theta_{avg} + \left(\frac{\sin \theta_{rcd} - \sin \theta_{adv}}{\pi} \right) \xi \qquad (2.93)$$

Substituting (2.93) into (2.92) and integrating, the surface tension component perpendicular to the substrate is obtained as

$$F_n = \pi \sigma r_b (\sin \theta_{adv} + \sin \theta_{rcd}) \qquad (2.94)$$

The gravity force component perpendicular to the substrate is

$$F_{g\perp} = \frac{\pi r_b^3 (2 - 3 \cos \theta_{avg} + \cos^3 \theta_{avg})}{3 \sin^3 \theta_{avg}} (\rho_1 - \rho_v) g \, \cos \alpha \qquad (2.95)$$

The maximum radius (r_{max}) of the drop that will initiate fall-off is obtained by balancing the forces perpendicular to substrate, namely

$$F_{g\perp} + F_n = 0 \qquad (2.96)$$

Hence, the critical radius of fall-off (r_{max}) is given as

$$r_{max} = \sqrt{ \left(\frac{3(\sin \theta_{avg})(\sin \theta_{rcd} + \sin \theta_{adv})}{(2 - 3 \cos \theta_{avg} + \cos^3 \theta_{avg})} \right) \left(\frac{\sigma}{(g \, \cos \sigma)(\rho_1 - \rho_v)} \right) } \qquad (2.97)$$

For $r_b > r_{max}$, the drop becomes unstable and falls-off.

2.4.6.6   Horizontal Substrate Having Unidirectional Wettability Gradient

A horizontal substrate with wettability gradient is shown in Fig. 2.23a. The contact angle at the lower wettability side is θ_1, while that at the higher wettability side is θ_2. The contact angle varies linearly in one direction from $x = 0$ to X. Here, X is the substrate length in x direction, Fig. 2.23b.

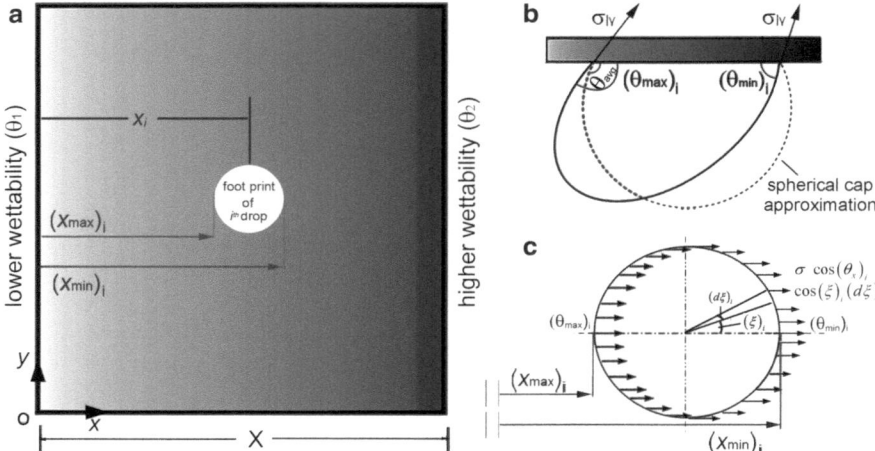

Fig. 2.23 Condensation over a substrate with wettability gradient. (**a**) The footprint of the *i*th drop is assumed to be circular. (**b**) Side view of each drop is determined by the two circle approximation. (**c**) Direction of forces acting over the three phase contact line—the substrate–vapor–liquid boundary

Consider the footprint of the *i*th drop of radius r at nucleation site (x_i, y_i) underneath a substrate having unidirectional wettability gradient. The side view of *i*th drop is shown in Fig. 2.23b. For intermediate calculations, the drop shape is taken to be a spherical cap in the sense that drop volume V_i, area of liquid–vapor interface $(A_{lv})_i$, and area of solid–liquid interface $(A_{sl})_i$ are calculated using the average contact angle

$$(\theta_{avg})_i = \frac{1}{2} [(\theta_{max})_i + (\theta_{min})_i] \tag{2.98}$$

Here, the contact angle of the drops depends on their position on the substrate. Consequently, force imbalance is generated primarily because of variation of contact angle from $(\theta_{max})_i$ to $(\theta_{min})_i$ at the three-phase contact, arising from the substrate wettability gradient, Fig. 2.23b. Drop motion can be expected even before the shapes are greatly altered by gravity or flow-related pressure nonuniformity. Gravity and pressure will not have component parallel to the horizontal substrate. The unbalanced surface tension will then mobilize the drop along the substrate.

The footprint of the spherical cap shape corresponding to the *i*th drop and as a circle is shown in Fig. 2.23c. It can be seen that the net force at the three-phase contact line of a deformed drop acts in the *x* direction towards the higher wettability side. It can be calculated as follows.

The base radius of the *i*th drop is

$$(r_b)_i = (r)_i \sin (\theta_{avg})_i \tag{2.99}$$

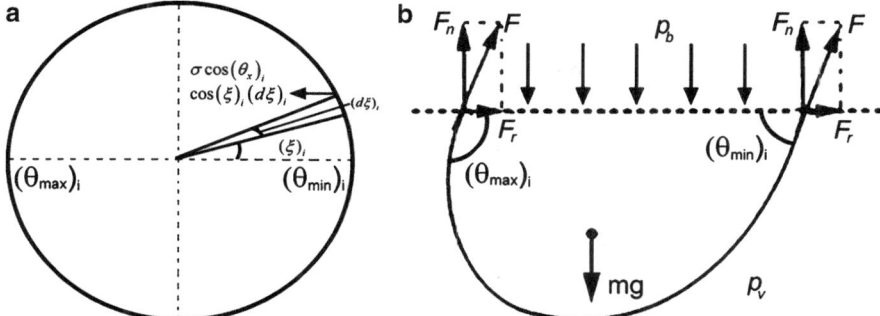

Fig. 2.24 (**a**) variation of contact angle with respect to the azimuthal angle at base of the drop is assumed to be a circle. (**b**) Free body diagram of ith drop underneath wettability gradient horizontal substrate

Quantities $x_{\min}$ and $x_{\max}$ for the ith drop are

$$(x_{\max})_i = x_i - (r_b)_i \quad \text{and} \quad (x_{\min})_i = x_i + (r_b)_i \qquad (2.100)$$

Angles $(\theta_{\max})_i$ and $(\theta_{\min})_i$ are calculated as

$$(\theta_{\max})_i = \theta_1 + \left(\frac{\theta_2 - \theta_1}{X}\right)(x_{\max})_i \qquad (2.101)$$

$$(\theta_{\min})_i = \theta_1 + \left(\frac{\theta_2 - \theta_1}{X}\right)(x_{\min})_i \qquad (2.102)$$

The net force acting at the footprint of ith the drop (Fig. 2.24a) towards higher wettability side is

$$(F_r)_i = 2\sigma \int_0^\pi \cos(\theta_x)_i \cos(\xi)_i (r_b)_i (d\xi)_i \qquad (2.103)$$

The value of $\cos(\theta_x)_i$ is linearly interpolated as

$$\cos(\theta_x)_i = \cos(\theta_{\min})_i + \left(\frac{\cos(\theta_{\max})_i - \cos(\theta_{\min})_i}{(x_{\min})_i - (x_{\max})_i}\right)(x)_i \qquad (2.104)$$

Substituting (2.104) into (2.103) and integrating, the retention force parallel to substrate is obtained as

$$(F_r)_i = (4/\pi)\sigma(r_b)_i[\cos(\theta_{\min})_i - \cos(\theta_{\max})_i] \qquad (2.105)$$

The retention force parallel to the substrate towards the higher wettability side is balanced by wall shear between the drop and the substrate. This balance requires that the drop will slide at a constant speed. The estimation of terminal velocity of a drop is discussed in Sect. 2.4.7. Hence, on the wettability gradient substrate every drop becomes unstable. The sliding of the drop wipes off other drops that lie in its path, and its mass and volume change during the motion. If the weight of the drop is higher than the net retention force evaluated at the three-phase contact line normal to the surface, the drop will fall-off. The critical radius of the drop at fall-off is estimated as follows.

The surface tension component normal to substrate is calculated as

$$(F_n)_i = 2 \int_0^\pi \sigma \, \sin \, (\theta_x)_i (r_b)_i \, \mathrm{d}\xi \qquad (2.106)$$

The variation of the contact angle, with respect to azimuthal angle along the contact line is linearly interpolated as

$$\sin \, (\theta_x)_i = \sin \, (\theta_{min})_i + \left(\frac{\sin \, (\theta_{max})_i - \sin \, (\theta_{min})_i}{(x_{min})_i - (x_{max})_i} \right)(x)_i \qquad (2.107)$$

Substituting (2.106) into (2.107) and integrating, the surface tension component perpendicular to the substrate is

$$(F_n)_i = \pi\sigma(r_b)_i [\sin \, (\theta_{min})_i + \sin \, (\theta_{max})_i] \qquad (2.108)$$

The gravity force component perpendicular to the substrate is

$$(F_{g\perp})_i = \frac{\pi(r_b^3)_i [2 - 3 \, \cos \, (\theta_{avg})_i + \cos^3(\theta_{avg})_i]}{3 \sin^3 \, (\theta_{avg})_i} (\rho_1 - \rho_v)g \qquad (2.109)$$

If the weight of the drop is higher than the net retention force evaluated at the three-phase contact line in a direction normal to the surface, the drop will fall-off. The corresponding critical radius is given by

$$(r_{max})_i = \sqrt{\frac{3\sigma \, \sin \, (\theta_{avg})_i [\sin \, (\theta_{avg})_i + \sin \, (\theta_{min})_i}{(\rho_1 - \rho_v)g[2 - 3 \, \cos \, (\theta_{avg})_i + \cos^3(\theta_{avg})_i]}} \qquad (2.110)$$

2.4.7 Modeling Terminal Velocity

In dropwise condensation, droplets undergo instability and start sliding over the substrate that is either inclined or is horizontal with a wettability gradient. The speed increases with time till the unbalanced force is matched by wall shear,

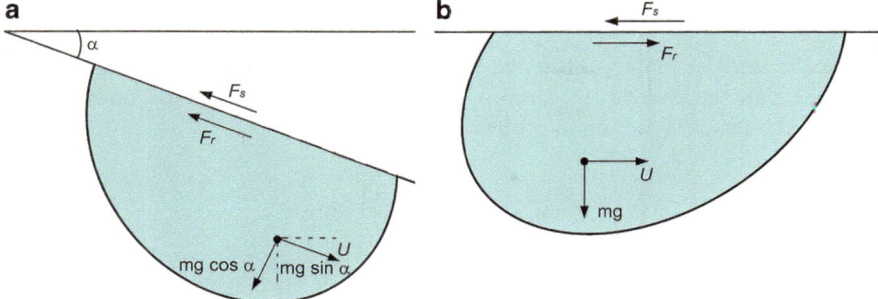

Fig. 2.25 Representation of various forces on sliding drop underneath substrate with terminal velocity (**a**) inclined substrate and (**b**) horizontal substrate with wettability gradient. The average skin friction coefficient (s_f) is determined at the scale of the individual drop by a CFD model described by the authors in their earlier study

resulting in a constant terminal velocity. In the present model it is assumed that drops attain terminal velocity immediately after instability.

2.4.7.1   Inclined Substrate

For an inclined substrate, the drop achieves terminal velocity when the component of weight parallel to the surface, retention force of the deformed drop at the three-phase contact line owing to surface tension, and wall friction are in balance, Fig. 2.25. Hence

$$F_{g\|} + F_{r\|} + F_s = 0 \tag{2.111}$$

where, $F_{g\|}$ is the component of weight parallel to the inclined substrate, $F_{r\|}$ is retention force opposing drop motion and F_s is wall shear associated with the relative velocity between the fluid and the substrate. The viscous force acting between the wall and fluid is obtained as follows

$$F_s = \frac{1}{2} C_f A_{sl} \rho_l U^2 \tag{2.112}$$

The terminal velocity of the drop over an inclined surface that makes an angle θ with the horizontal can now be calculated as

$$U \doteq \sqrt{\frac{2(F_{g\|} - F_r)}{C_f \rho_l A_{sl}}} \tag{2.113}$$

2.4.7.2 Horizontal Substrate with a Wettability Gradient

For an inclined substrate, when the component of surface tension parallel to it balances wall shear stress, the drop will slide with constant speed. The hydrodynamic force that resists motion of the ith drop is

$$(F_{\text{hyd}})_i = C_{\text{f}}\left(\frac{1}{2}\rho U_i^2\right)(A_{\text{sl}})_i \qquad (2.114)$$

The skin coefficient of friction C_{f} is obtained from the correlation derived from the CFD model described in Chap. 3. Equating the expressions of forces given by (2.105 and 2.114)

$$(F_{\text{r}})_i + (F_{\text{hyd}})_i = 0 \qquad (2.115)$$

The terminal velocity is obtained as

$$U_i = \left[\frac{0.022(F_{\text{r}})_i(\theta_{\text{avg}})_i^{1.58}}{\rho^{0.03}\mu^{0.97}(r_{\text{b}}^{1.03})_i}\right]^{1/1.03} \qquad (2.116)$$

Velocity thus obtained is a function of the drop size and its position for a wettability gradient surface.

2.4.8 Wall Heat Transfer

Heat transfer during dropwise condensation can be calculated from the rate of condensation at the free surface of the drop at each nucleation site of the substrate. The gaps between drops are assumed to be inactive for heat transfer. The heat transfer rate q is a function of the nucleation site density and the rate of growth of drop radius at each nucleation site. The latter is estimated by using a quasi-one-dimensional approximation for thermal resistances, including the interfacial and capillary resistance at the vapor–liquid boundary and conduction resistance through drop, as discussed in Sect. 2.4.4.1 and given by (2.62). The rate of condensation of vapor at each nucleation site can now be determined as follows.

Estimate the number of available nucleation sites (N). The mass of condensate accumulated at the ith nucleation site over a time interval Δt is

$$(\Delta m)_i = \rho\frac{\pi}{3}(2 - 3\cos\theta_{\text{avg}} + \cos^3\theta_{\text{avg}})(r_{\text{new}}^3 - r_{\text{old}}^3)_i \qquad (2.117)$$

With N, the number of active nucleation sites, the total quantity of condensate at a given time step (Δt) is obtained as

$$\Delta m = \sum_{i=1}^{i=N} (\Delta m)_i \qquad (2.118)$$

Therefore, the average rate of condensation underneath a substrate is given by

$$(m)_{\text{avg}} = \frac{1}{t} \left(\sum_{j=1}^{j=K} (\Delta m)_j \right) \quad \text{where,} \quad t = \left(\sum_{j}^{j=K} (\Delta t)_j \right) \qquad (2.119)$$

Here, t is the time period of condensation and $K = (t/\Delta t)$ is the number of time steps. The heat transfer rate is simply the latent heat released during the condensation divided by the time elapsed. The sensible cooling of the liquid is neglected. The average heat transfer coefficient over an area (A) of the substrate during dropwise condensation is then determined from the formula

$$h = \frac{m_{\text{avg}} h_{\text{lv}}}{A(T_{\text{sat}} - T_{\text{w}})} \qquad (2.120)$$

2.4.9 Area of Coverage

At each time step, the number of available nucleation sites and size of the drop at these locations is obtained from simulation. The area of substrate covered by the condensate is determined as follows:

N is the available nucleation sites at a given time t. The area of coverage at the ith nucleation site is calculated as

$$(A_{\text{sl}})_i = \pi r_i^2 (1 - \cos^2 \theta_{\text{avg}}) \qquad (2.121)$$

With N and A, the number of active nucleation sites and the total substrate area, the area covered by drops at a given time is obtained as

$$A_{\text{cd}} = \frac{1}{A} \left(\sum_{i=1}^{i=N} (A_{\text{sl}})_i \right) \qquad (2.122)$$

Here, A_{cd} is the fraction of area of substrate covered by drops at a given time instant. With K, the number of time instants within a cycle of time period t, the average percentage of area covered for a cycle is as

$$\frac{1}{A} \left(\frac{1}{t} \sum_{j=1}^{K} A_{\text{cd}} \right) \times 100 \qquad (2.123)$$

2.4.10 Available Liquid–Vapor Interface Area

Condensation takes place only over the liquid–vapor interface of the drop. The area of liquid–vapor interface at the ith nucleation site drop at a given time is calculated as

$$(A_{lv})_i = 2\pi r_i^2(1 - \cos\theta_{avg}) \tag{2.124}$$

With N, the number of active nucleation sites, the fraction of liquid–vapor interface area available for condensation on a substrate of area A at a given time instant is

$$A_{alv} = \frac{1}{A}\left(\sum_{i=1}^{i=N}(A_{lv})_i\right) \tag{2.125}$$

2.5 Numerical Algorithm for Dropwise Condensation

The modeling of dropwise condensation considers the details of each subprocess of the condensation cycle and interrelates them in such way as to form a full cycle. The important steps in the numerical algorithm can now be stated as follows:

1. Initialize all the variables such as thermophysical properties, physicochemical properties of the substrate, type of substrate (with or without wettability gradient), orientation of substrate, nucleation site density, time-step, and total time of simulation.
2. Distribute the nucleation sites on the substrate using a random number generator and place the drop of minimum radius at all the nucleation sites.
3. Solve (2.62) by a fourth order Runge-Kutta method over a time step and find the new radius.
4. Calculate the distance between nucleation sites.
5. Check for the coalescence.
6. Identify the nucleation sites covered by the resulting coalesced drops and keep them deactivated till the drop covers them.
7. Simultaneously, search for newly exposed sites created due to drop coalescence and provide a minimum radius drop on such newly exposed sites.
8. For all drops, check for the critical radius of slide-off and the sliding velocity.
9. Reactivate the exposed sites created due to drop slide-off and provide a minimum radius drop on newly exposed sites.
10. Check for drop fall-off.
11. Reactivate the exposed sites created due to drop fall-off and provide a minimum radius drop on the newly exposed sites.
12. Repeat (3)–(9) till a dynamic steady-state is reached.

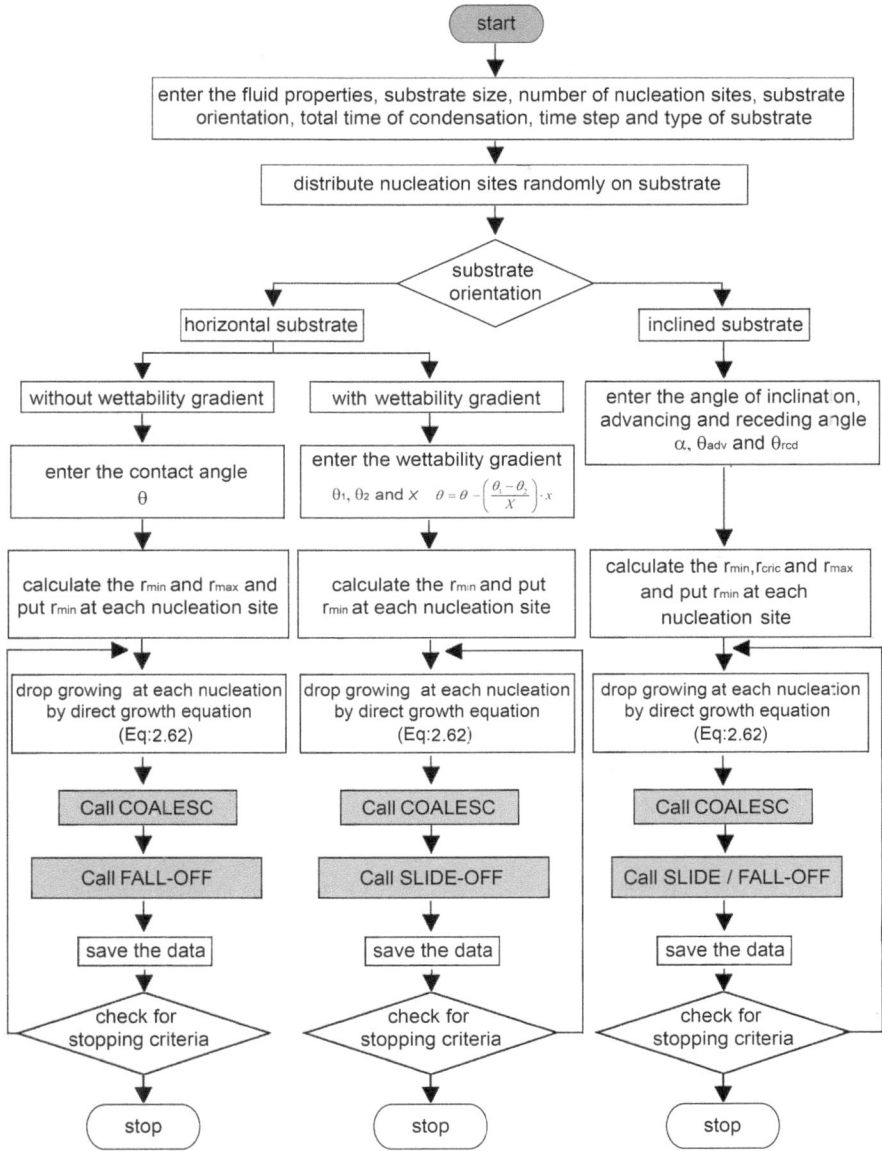

Fig. 2.26 Flow diagram of the dropwise condensation model

A computer program in C++ is written to carry out simulation as per the proposed algorithm. It is run on a high performance computing machine. The flow chart of the mathematical model of dropwise condensation underneath the inclined substrates is depicted in Fig. 2.26, while the subroutines are detailed in

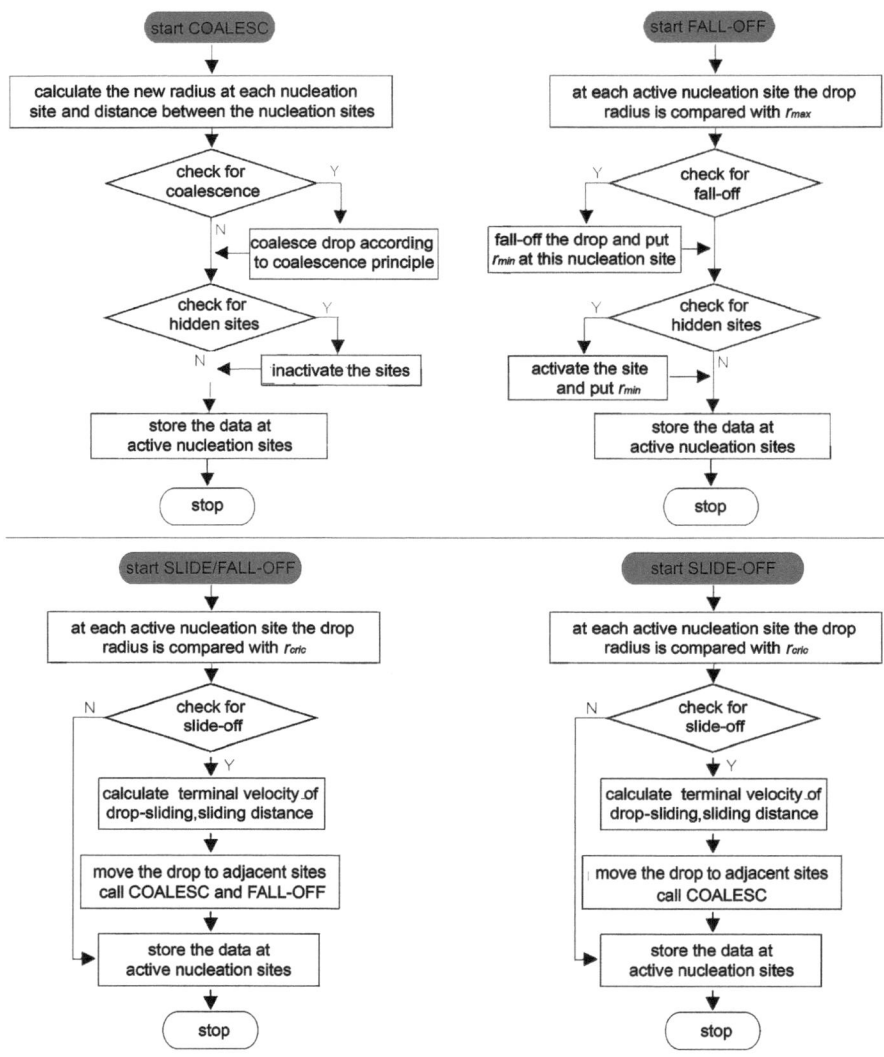

Fig. 2.27 Various subroutines of the dropwise condensation model

Fig. 2.27. Whenever a drop is removed or shifted from its location due to sliding and coalescence all the hidden nucleation sites underneath the drop become active and are immediately supplied with thermodynamically stable droplets of the minimum radius. It is to be noted that the simulation needs to track multiple generations of the droplets—nucleating, growing by direct condensation, by coalescence and some slide/falling-off—when the virgin surface thus exposed is re-nucleated. The computations are hence, quite intensive.

While the mathematical model developed is quite general, simulations have been carried out under the following assumptions:

1. Nucleation sites are randomly distributed on the surface. Unless stated otherwise, all computations have been performed with an initial nucleation site density of $10^6 \, \text{cm}^{-2}$.
2. Thermodynamically constrained smallest radius is taken as the minimum radius in the simulation. Initially, the substrate is dry and all the nucleation sites are instantaneously occupied by the droplet of minimum radius.
3. Heat transfer resistance arises due to the liquid–vapor interface, curvature, and conduction, driven by imposed subcooling of the substrate. Convective transport of heat is neglected for static drops but is included for a drop in motion. Constriction resistance is neglected.
4. The accommodation coefficient is taken to be 0.035 for water, 0.45 for mercury, and 0.21 for sodium and potassium (Carey 2008).
5. Droplet coalescence is assumed to be instantaneous and resulting droplet attains instantaneous mechanical stability; interface oscillations are neglected. Also, change in the shape of the drop due to acceleration is neglected.
6. An equivalent spherical-cap approximation has been incorporated to model the droplet shapes. For drops on inclined surfaces, the two-circle approximation is used (Elsherbine and Jacobi 2004a, b).
7. Though contact angles are obtained (from theory or experiments) under static conditions, these values have been used under dynamic conditions as well.
8. Partial fall-off of the drops is neglected in the sense that instability results in the complete volume of the drop being removed.
9. The entire substrate is assumed to be at a constant temperature; drop motion leads to changes in the wall heat flux; local wall temperature fluctuations observed by Bansal et al. (2009) have been neglected.
10. Thermophysical properties of the vapor and liquid phases are taken to be independent of the temperature; the vapor is saturated; all the properties are evaluated at the average of the substrate and saturation temperatures.

2.6 Substrate Leaching

A consequence of the time-dependent processes in dropwise condensation associated with the movement of the drop, first by coalescence and then by sliding motion, is to reduce sustainability on or underneath an inclined chemically textured substrate. Hence, the life of a condensing surface depends on the wall shear interaction of sliding droplets with the drop promoter layer. The phenomenon of removal of the promoter layer over the substrate is called surface leaching. It arises primarily from viscous forces at the contact surface and chemical reactions between the condensing liquid and the promoter. Heat transfer rates and temperature fluctuations affect these interactions. Accordingly, the long-term sustainability of the process is greatly reduced. Hence, coalescence and sliding of drop in dropwise condensation are significant for improving heat transfer coefficient but reduce the substrate life. Even if there is no chemical reaction between the promoter and condensing liquid, the wall shear stress becomes the primary quantity that controls leaching. A prediction of shear stress requires a complete knowledge of the flow field inside the droplets during coalescence and sliding. Given a shear stress distribution for an individual drop, the net effect due to a drop ensemble can be determined from the time-averaged drop size distribution.

Literature on surface leaching due to drop motion is limited. Therefore, a detailed simulation of flow and heat transfer in a liquid drop sliding underneath a hydrophobic surface and determination of local distribution wall shear stress and wall heat transfer of individual drop form one of the motivations of the present study.

2.7   Closure

A comprehensive mathematical model of dropwise condensation underneath an inclined substrate with and without wettability gradient is presented. The dropwise condensation process is hierarchical because it starts from the atomic scale and progresses on to the engineering scale. The mathematical models of various subprocesses in dropwise condensation have been reported and these are interrelated according to the experimental observations. The overall flow chart of simulation is shown in Fig. 2.26 while Fig. 2.27 shows various subroutines. A C++ program is written to carry out model simulations.

Chapter 3
Dropwise Condensation: Simulation Results

Keywords Numerical simulation • Condensation patterns • Area coverage • Area-averaged heat transfer coefficient • Surface inclination • Drop size at criticality • Cycle time

3.1 Dropwise Condensation of Water Vapor

After validation, simulations have been performed for water vapor condensation underneath a horizontal and an inclined textured substrate. A horizontal surface having unidirectional wettability gradient has also been considered. Here, the effect of thermophysical properties, physicochemical properties of the substrate, promoter layer thickness, nucleation site density, saturation temperature, degree of subcooling, effect of wettability gradient and angle of inclination are parametrically explored. Unless otherwise stated, we have used nucleation site density of $10^6 \, \text{cm}^{-2}$ in the simulations reported here.

3.1.1 Effect of Substrate Hydrophobicity

The effect of the hydrophobicity of the substrate towards the condensing liquid is examined in Figs. 3.1, 3.2, and 3.3. Here, drop distribution at selected time instants is pictorially depicted from initial nucleation to the first instance of drop fall-off. Water vapor condenses at 303 K underneath the surface and the degree of subcooling is 5 K. The substrate is horizontal and various contact angles are considered.

A time sequence of condensation patterns for a contact angle of 90° is shown in Fig. 3.1. Drop diameter at criticality is 4.63 mm and fall-off first occurs at 50.15 min after commencement of condensation. The corresponding spatiotemporal drop distribution for a contact angle of 120° is shown in Fig. 3.2. Drop diameter at criticality is 3.088 mm and fall-off first occurs at 21.55 min after commencement of

S. Khandekar and K. Muralidhar, *Dropwise Condensation on Inclined Textured Surfaces*, SpringerBriefs in Applied Sciences and Technology ·11, DOI 10.1007/978-1-4614-8447-9_3, © Springer Science+Business Media New York 2014

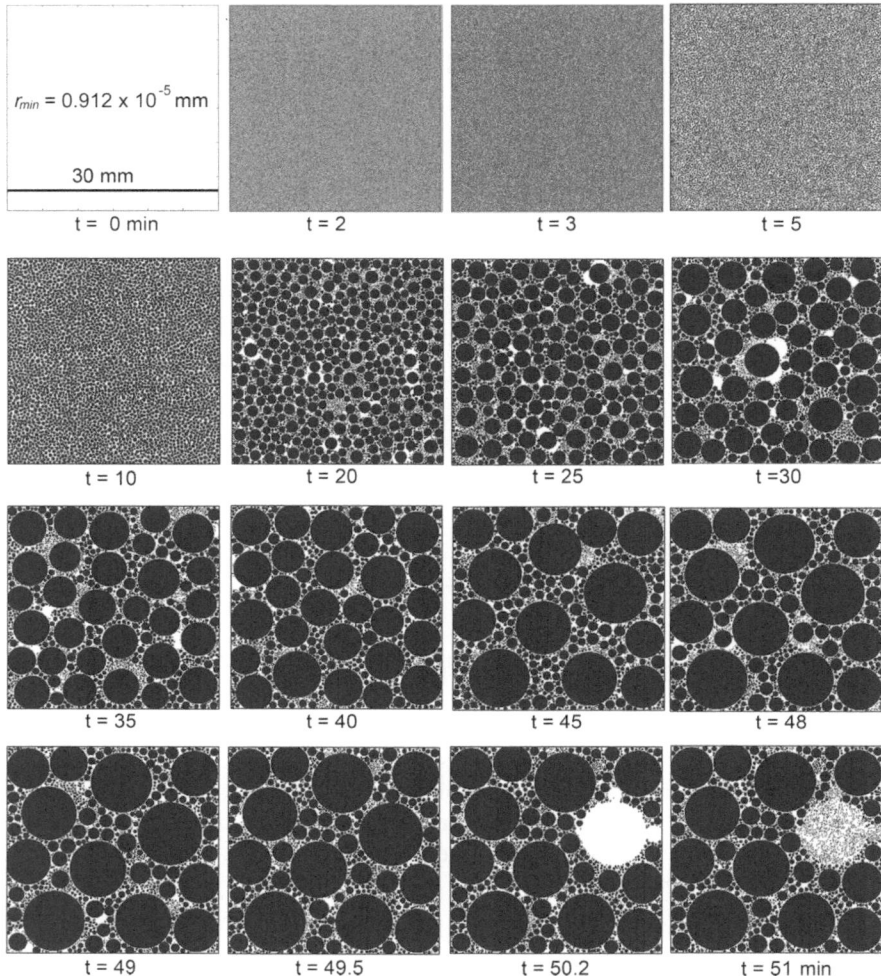

Fig. 3.1 Drop distribution from the start to the first fall-off during dropwise condensation of water vapor at 303 K with subcooling of $\Delta T = 5$ K underneath a horizontal substrate of contact angle 90°

condensation. The spatiotemporal drop distribution for a contact angle of 140° is shown in Fig. 3.3. Drop diameter at criticality is 2.14 mm and fall-off first occurs at 7.25 min after commencement of condensation.

A reduction in wettability increases the contact angle and leads to a smaller base circle of the drop and, therefore, smaller surface forces retaining the drop against gravity. Thus, two effects are clearly visible. (1) The droplet volume at the time of fall-off is smaller, Figs. 3.1, 3.2, and 3.3. (2) With increasing contact angle, the drops achieve fall-off criticality earlier in the cycle.

The area of coverage by drops for various contact angles is shown in Fig. 3.4a. The hydrophobicity of substrates decreases the area of coverage. Therefore, highly

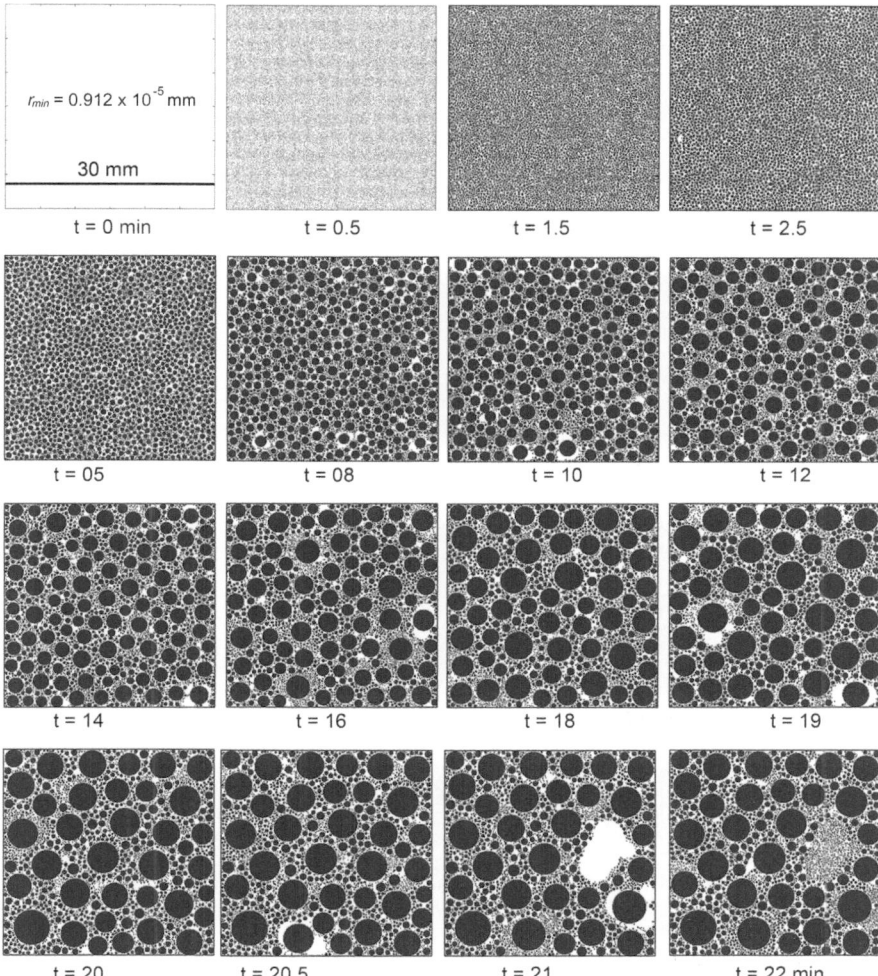

Fig. 3.2 Drop distribution from the start to the first fall-off during dropwise condensation of water vapor at 303 K with subcooling of $\Delta T = 5$ K underneath a horizontal substrate of contact angle 120°

hydrophobic substrates (higher contact angle) have higher available nucleation sites density at any given time of condensation, Fig. 3.4b. The size and population of maximum diameter drops have a significant impact on dropwise condensation, due to the limitations imposed by the diffusional resistance of the liquid. The effect of substrate hydrophobicity on the surface-averaged heat transfer rate during dropwise condensation is shown in Fig. 3.4c. The apparent contact angles clearly show an effect on heat transfer. It is clear that the size of the drop at fall-off as well as the time required for fall-off decrease as the hydrophobicity of the substrate increases. All other conditions remaining unchanged, the fall-off time for a pendant drop is seen to be a linear function of the contact angle, as shown in Fig. 3.4d.

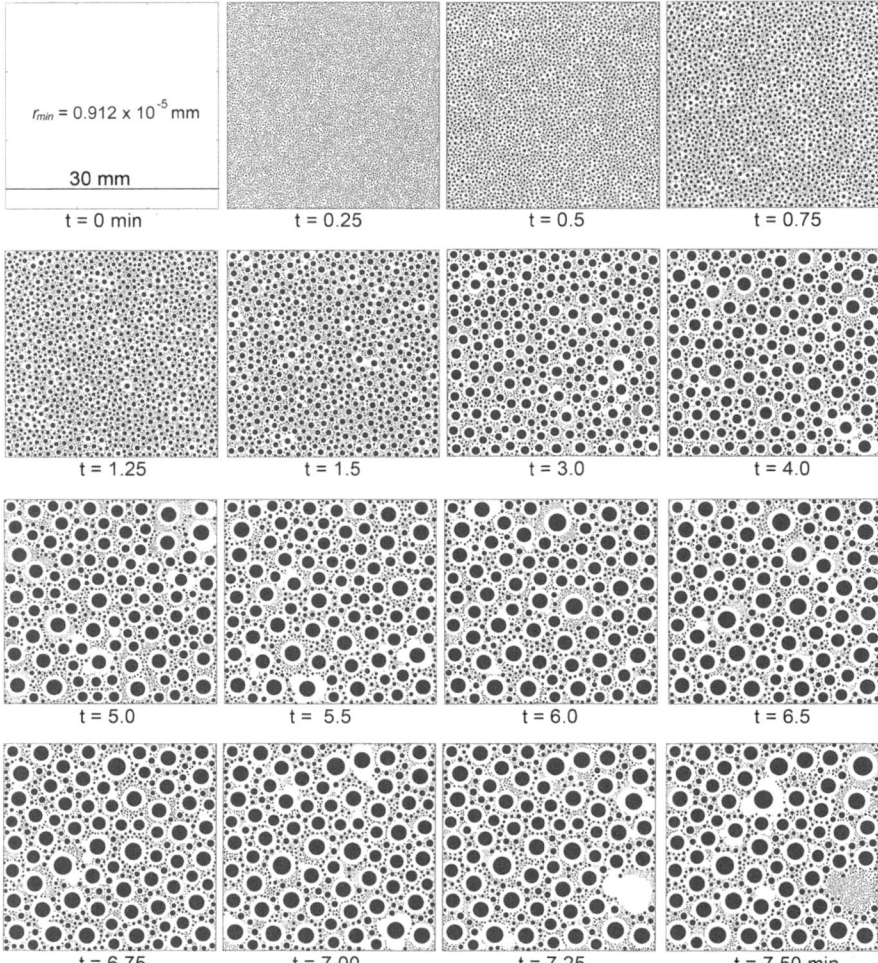

Fig. 3.3 Drop distribution from the start to the first fall-off during dropwise condensation of water vapor at 303 K with subcooling of $\Delta T = 5$ K underneath a horizontal substrate of contact angle 140°

For a given initial nucleation site density on a substrate, heat transfer can be increased by having a contact angle higher than 90°, i.e., making the substrate hydrophobic. Accordingly, substrates having higher hydrophobicity result in the condensate having drop size distributions towards the smaller diameter (Figs. 3.1, 3.2, and 3.3), and vice-versa. It results in lowering the overall diffusional resistance to heast transfer which is offered by the condensing drops. In addition, increased hydrophobicity generates a large number of nucleation sites, at any given time.

The nucleation sites available for nucleation are shown in Fig. 3.4b. Initially, it the number of nucleation sites decrease according to a power law but after reaching a

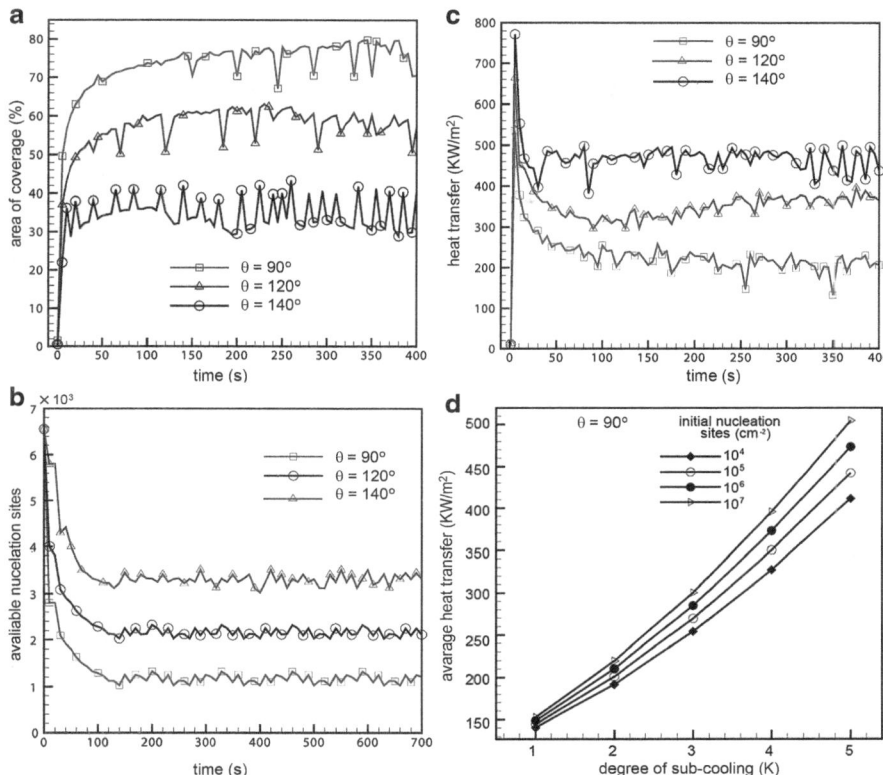

Fig. 3.4 (**a**) Area of coverage as a function of time, (**b**) available nucleation sites over the substrate of size 30 mm × 30 mm with respect to time for various contact angles at 303 K and degree of subcooling = 5 K, (**c**) fluctuations in heat transfer rate on a substrate with respect to time and (**d**) fall-off time of a drop as function of the contact angle

dynamic steady-state it varies quasi periodically due to coalescence and fall-off of droplets. The frequency of drop fall-off, size of the minimum drop and size of the maximum drops which gets formed, for substrates having various degree of surface hydrophobicity, during condensation of water vapor on it are summarized in Table 3.1.

As given in Table 3.1, the substrate hydrophobicity decreases the critical size (of fall-off), hence resulting in a higher population of small drops. Therefore, one concludes that a substrate having a high hydrophobicity with the condensate fluid is desirable in dropwise condensation. The frequency (namely, the number of drops) as a function of the drop radius, 10 min after commencement of condensation for a contact angle of 90°, saturation temperature = 303 K and subcooling = 5 K, is shwon in Fig. 3.5a. At later times, drops of higher sizes are to be seen. For the present simulation, the fall-off time of the first drop was approximately 50.2 min. Very small droplets nucleate on the substrate at 50 min, immediately before the first drop falls-off at 50.2 min, is shown in Fig. 3.5b.

Table 3.1 Results summarizing parameters of dropwise condensation of water vapor at 303 K and degree of subcooling $= 5$ K after reaching a quasi-steady state

Contact angle (°)	Radius r_{min} (mm)	r_{max} (mm)	Initial nucleation sites (cm^{-2})	Available nucleation sites (30 mm × 30 mm)	First fall-off (min)	Cycle time (s)	Heat transfer (kW/m^2)
90	9.1×10^{-6}	4.64	10^6	1,122	50.2	360	270
			10^7	1,137	38.5	187	282
120	9.1×10^{-6}	3.08	10^6	1,977	21.5	242	395
			10^7	2,015	17.8	155	512
140	9.1×10^{-6}	2.14	10^6	3,656	7.2	189	525
			10^7	3,684	6.7	85	580

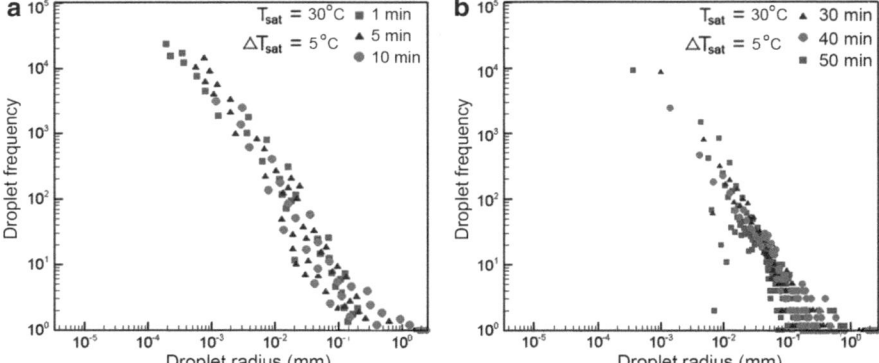

Fig. 3.5 (**a, b**) Temporal variation in drop size distribution for condensing water vapor underneath a horizontal substrate for contact angle 90°. For clarity, data for 1–10 min are separately plotted from the data 30–50 min. The fall-off time for the first drop was equal to 50.2 min in this simulation

3.1.2 Effect of Substrate Inclination

The distribution of drops arising from water vapor condensation at 303 K and degree of subcooling $= 5$ K from initial nucleation to the first slide-off underneath substrates of various orientation (30°, 60°, and 90°) is shwon in Figs. 3.6, 3.7, and 3.8. The pictorial views of condensations have site density 10^6 cm^{-2}, substrate size 30 mm × 30 mm for substrates of inclination 30° and 60° and 20 mm × 20 mm for an inclination of 90°. For these simulations, the advancing and receding angles are taken as 118.5° and 101.5° yielding an average contact angle of 110° and a contact angle hysteresis of 17°.

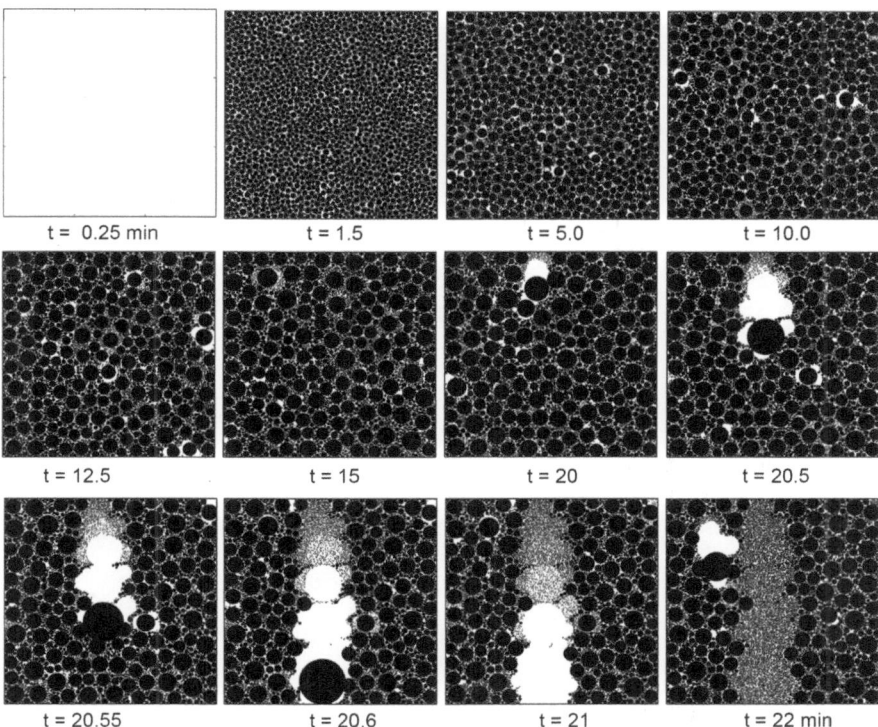

t = 0.25 min	t = 1.5	t = 5.0	t = 10.0
t = 12.5	t = 15	t = 20	t = 20.5
t = 20.55	t = 20.6	t = 21	t = 22 min

Fig. 3.6 Drop distribution from the start to the first slide-off during dropwise condensation of water vapor at 303 K with subcooling of $\Delta T = 5$ K underneath an inclined substrate. The angle of inclination is 30° and the size of the substrate is 30 mm × 30 mm

Condensation patterns on various inclinations show similarity. The point of difference is the size of the drop at slide-off and the average cycle time for slide-off from the substrate. These quantities decrease as the angle of inclination with respect to the horizontal increases.

The variation of the critical size of the droplet with respect to substrate orientation is depicted in Fig. 3.9. The critical drop size decreases with substrate hydrophobicity and inclination, causing a reduction in the cycle time and hence results in more frequent instances of re-nucleation. The reduction in the cycle-averaged drop size is an important factor in increasing heat transfer from strongly hydrophobic surfaces.

For ease of calculation, the data of Fig. 3.9 is correlated for various surfaces as

$$\frac{r_{\text{crit}}}{r_{\text{cap}}} = (2.1612 - 0.7699\,\theta_{\text{avg}})(\Delta\theta)^{0.5}\alpha^{-0.4266} \tag{3.1}$$

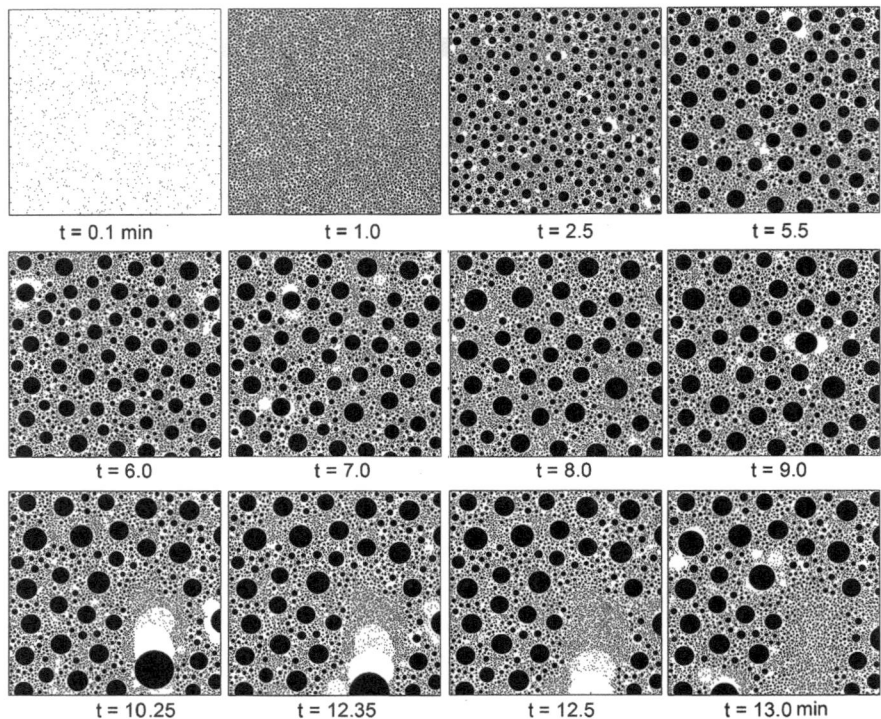

Fig. 3.7 Drop distribution from the start to the first slide-off during dropwise condensation of water vapor at 303 K with subcooling of $\Delta T = 5$ K underneath an inclined substrate. The angle of inclination is 60° and the size of the substrate is 30 mm × 30 mm

The correlation coefficient of (3.1) is 99.8 % as shown in Fig. 3.9b. Equation (3.1) simplifies dropwise condensation calculations within the hierarchical model and considerably reduces the computational time.

The heat transfer coefficient of dropwise condensation for a given saturation temperature and degree of subcooling is dependent on the contact angle, its hysteresis and substrate orientation, which in turn affect the criticality of sliding. The effect of critical radius on heat transfer at 303 K and degree of subcooling 5 K is depicted in Fig. 3.10a. It is seen that heat transfer coefficient increases as drop departure radius decreases, the correlation being

$$h = 0.19 \, r_{\text{crit}}^{-1.2} \tag{3.2}$$

Here, h is in unit of MW/m^2 K and r_{crit} is in mm.

The variation of heat transfer coefficient with respect to the angle of inclination of the substrate is shown in Fig. 3.10b. Numerically obtained data of water condensing underneath various inclined (0–90°) substrates is shown for an average contact angle of 110°, contact angle hysteresis of 17° and degree of

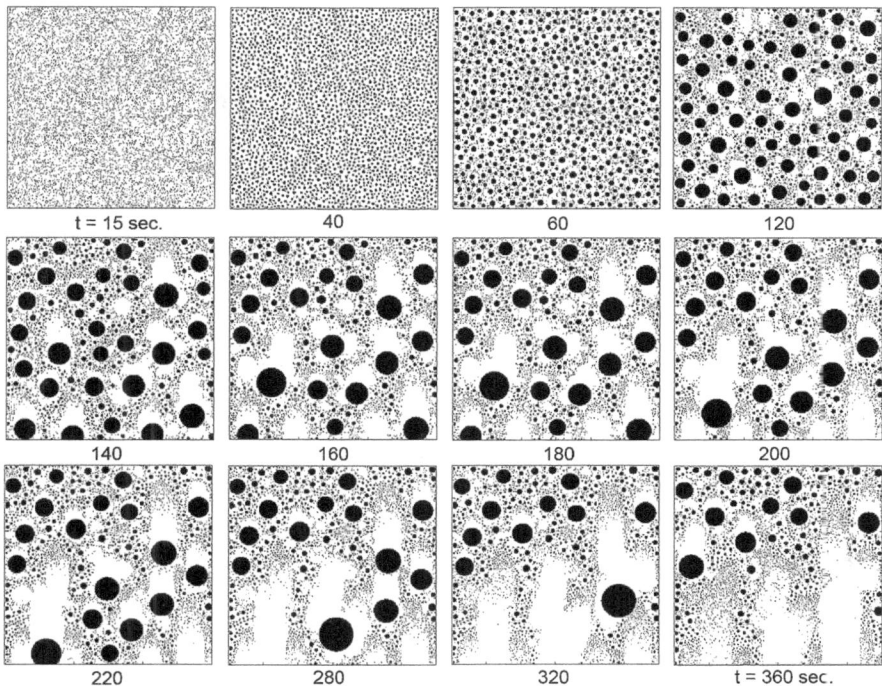

Fig. 3.8 Drop distribution from the start to the first slide-off during dropwise condensation of water vapor at 303 K with subcooling of $\Delta T = 5$ K underneath an inclined substrate. The angle of inclination is 90° and the size of substrate is 20 mm × 20 mm

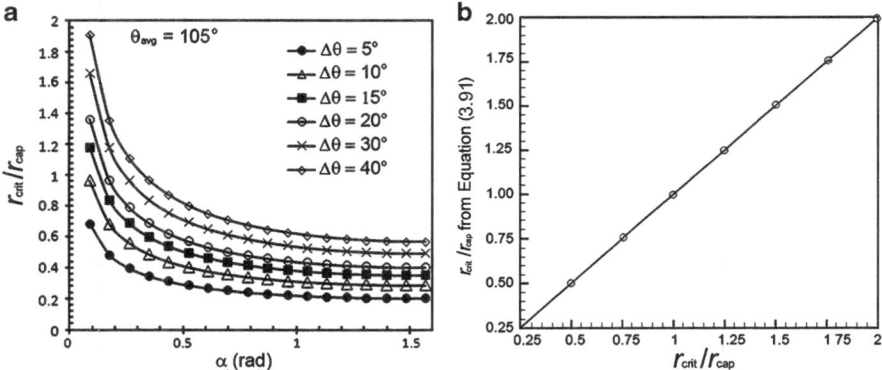

Fig. 3.9 (a) Critical base radius of the drop at instability (r_{crit}/r_{cap}) for a pendant arrangement plotted as a function of substrate orientation (α). Contact angle hysteresis is a parameter while the average contact angle is 105°. (b) Parity plot between (a) and the correlation, 3.2, show an excellent match

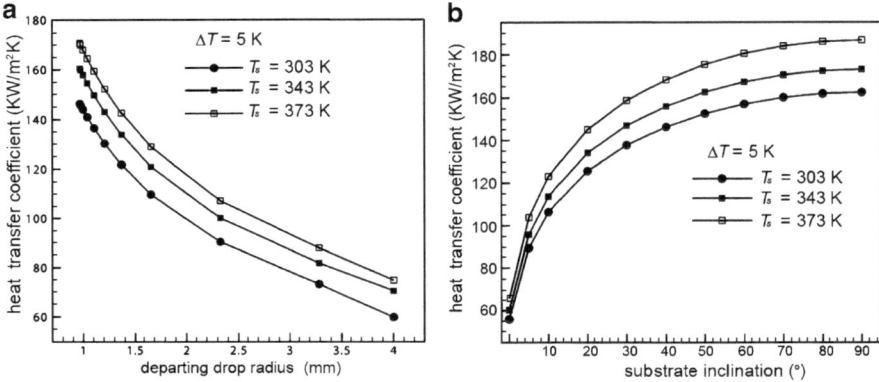

Fig. 3.10 (**a**) Dependence of heat transfer on the departure drop radius. (**b**) Effect of substrate inclination on heat transfer coefficient

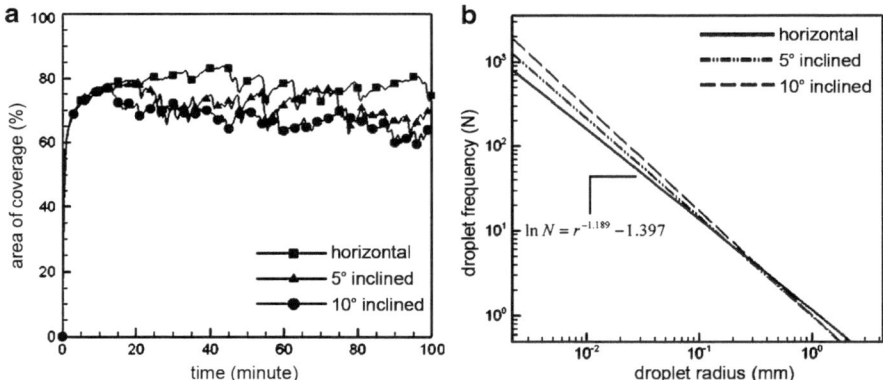

Fig. 3.11 Effect of substrate inclination: (**a**) temporal variation of area coverage of drops during condensation of water in the pendant mode. (**b**) Drop size distribution just before fall-off (for horizontal substrate) or slide-off (inclined substrate). For this simulation, $\theta_{adv} = 118.5°$ and $\theta_{rcd} = 101.5°$. $T_s = 303$ K, $\Delta T = 5$ K

subcooling = 5 K at various saturation temperatures. The result exhibits a 40–50 % higher heat transfer coefficient for vertical substrate as compared to the horizontal substrate.

The effect of substrate inclination on the temporal distribution of area coverage of drops is presented in Fig. 3.11a. Inclination of the substrate facilitates easier movement of drops by sliding, leading to a sweeping action. Therefore, the effective steady state coverage is smaller for inclined substrates, changing from 76.1 % for a horizontal substrate, 71.2 % for 5° inclination and 67.4 % for the substrate with 10° inclination. At the instant of the first fall-off (for the horizontal substrate) and the first slide-off (for the inclined substrate), Fig. 3.11b depicts the drop size

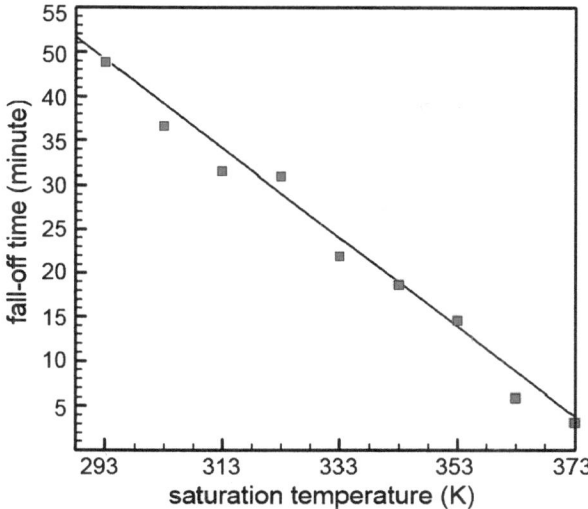

Fig. 3.12 Variation in drop departure time (time required for first fall-off) on a horizontal substrate with respect to the saturation temperature. Fluid employed is water, subcooling ΔT_{sat} = 5 K, contact angle = 110°, nucleation site density = 10^6 cm^{-2}. For a given nucleation site density, the fall time has an uncertainty of ±3 min, depending on the random assignment of initial droplet centers on the substrate

distribution as a function of radius, for various inclination angles. The distribution follows a power law with the negative slope increasing with angle, reflecting the repeated appearance of small drops at fresh nucleation.

3.1.3 Effect of T_{sat} and Subcooling

For a given degree of subcooling (ΔT_{sat} = 5 K), the effect of saturation temperature on drop departure time underneath a horizontal substrate is shown in Fig. 3.12. Increasing the saturation temperature reduces the fall-off time and hence the size of the largest drop, indicating an increase in the overall heat transfer coefficient. The diffusional thermal resistance within the drop is a major limiting factor of condensation heat transfer. Hence, increasing the saturation temperature increases the thermal conductance of the water drop. A marginal increase in the overall resistance is also noticed due to a reduction in the interfacial heat transfer coefficient; it essentially proves to be inconsequential as the overall thermal resistance is dominated by conduction resistance of the droplet.

The effect of saturation temperature and degree of subcooling on the heat transfer coefficient of dropwise condensation underneath horizontal and vertical substrates, is shown in Fig. 3.13. The data show a tendency of increasing heat transfer

Fig. 3.13 Dependence of
heat transfer coefficient on
saturation temperature and
degree of subcooling.
(**a**) Horizontal substrate,
(**b**) vertical substrate.
Condensing fluid is water
average, contact angle is
110°, contact angle
hysteresis is 17° and
nucleation sites density
is 10^6 cm^{-2}

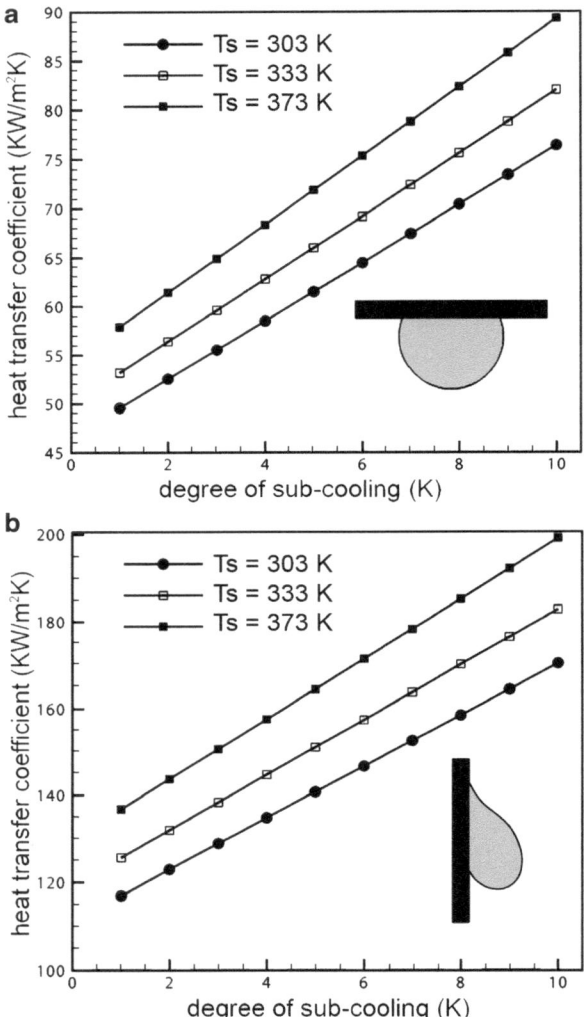

Fig. 3.13 Dependence of heat transfer coefficient on saturation temperature and degree of subcooling. (**a**) Horizontal substrate, (**b**) vertical substrate. Condensing fluid is water average, contact angle is 110°, contact angle hysteresis is 17° and nucleation sites density is 10^6 cm^{-2}

coefficient with increasing saturation temperature and degree of subcooling. This is
caused mainly by the decrease in the interfacial resistance at high saturation temperature of water vapor.

Increasing the degree of subcooling increases the density of active nucleation
sites on the condensing substrate. On the basis of numerical data, heat transfer
coefficient (kW/m^2 K) is empirically correlated with the critical radius of drop
(mm), degree of subcooling and saturation temperature (both in units of °C) for
water vapor condensation underneath an inclined substrate with a nucleation site
density of 10^6 cm^{-2} as follows:

$$h = (0.42\,\Delta T + 6.4)T_{sat}^{0.75}\,r_{crit}^{-1.18} \tag{3.3}$$

Here, the critical radius depends on contact angle, its hysteresis and angle of inclination of substrate from horizontal. It is preferable to cast this correlation in dimensionless form, applicable for all the inclinations, saturation temperature, and degree of subcooling. The dimensionless heat transfer correlation for water vapor condensation underneath an inclined substrate is obtained as

$$\text{Water(H}_2\text{O}) : \text{Nu} = (8.54 \times 10^3\,\text{Ja} + 240)\left(\frac{T_{sat}}{T_{ref}}\right)^{0.75}\left(\frac{r_{crit}}{r_{cap}}\right)^{-1.18} \tag{3.4}$$

Equation (3.4) has a correlation coefficient of 98.5 %. Standard reference values are used are properties of water at the normal boiling point.

3.1.4 Effect of Nucleation Site Density

Increasing the density of nucleation sites leads to a large overall heat transfer in dropwise condensation. This effect arises mainly from a reduction in the size of the drop before coalescence. Early coalescence allows virgin spaces for new initial drops, causing a high population of small drops.

One can conclude that a surface providing a higher nucleation sites is desirable for dropwise condensation. The number of nucleation sites is chosen as a parameter for condensation underneath a horizontal substrate, the working fluid being water. The effect of initial nucleation site density on heat transfer is shown in Fig. 3.14 for a contact angle of 110° and a saturation temperature of 303 K.

As the number of nucleation sites increases per unit area, many small drops nucleate on exposure of the surface to vapor, i.e., the average drop size within a cycle decreases. The conduction resistance is thus lowered, leading to an increase in average heat transfer coefficient.

3.1.5 Effect of Promoter Layer Thickness

Dropwise condensation of water underneath metal surfaces is rarely observed in natural conditions. It is generally promoted with a suitable coating. An understanding of the role of coatings is critical not only because it determines the surface wettability but it adds an extra thermal resistance. The nucleation density is also dependent on the promoter layer. The mathematical model of the present work is utilized for designing and quantifying the effect of the coating.

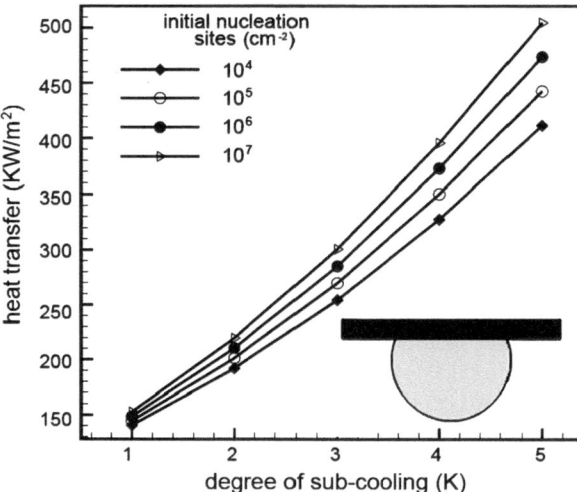

Fig. 3.14 Overall heat transfer rate as a function of subcooling with nucleation site density as a parameter

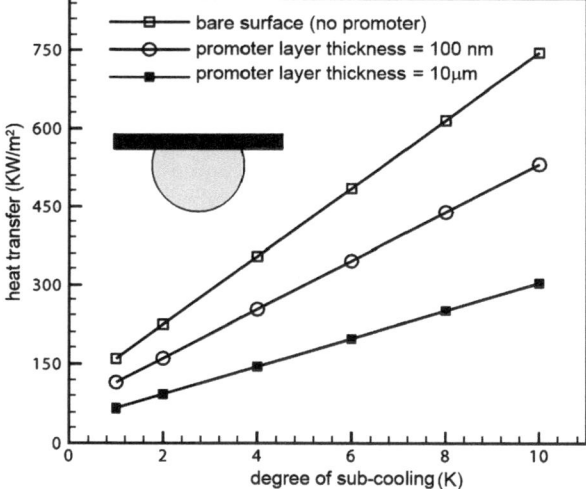

Fig. 3.15 Overall heat transfer as a function of the degree of subcooling with varying thickness of the promoter layer underneath a horizontal substrate. Fluid considered is water at 303 K, contact angle of 100°, and nucleation site density $= 10^6$ cm^{-2}

The overall heat flux for coatings of varying thicknesses is presented in Fig. 3.15. The computation was carried out when the layer with the thermal conductivity of 0.28 W/m K creates a 100° contact angle.

The overall heat flux is significantly influenced by the thickness. When the 10 μm thick promoter thickness is reduced to 100 nm without any change in the contact angle and nucleation density, heat transfer improves by a factor of 1.75. If the same hydrophobicity can be achieved without any promoter, the condensing surface can produce 1.4 times the heat transfer of the 100 nm thick promoter and 2.45 times that of the 1 μm promoter, respectively. This result shows that a redundantly thick coating results in a significant degradation of heat transfer.

3.1.6   Effect of Wettability Gradient

The effectiveness of dropwise condensation is improved by moving the liquid drop that grows on or underneath a solid substrate. The droplet moving over the surface wipes other droplets off. Consequently, more unexposed area is available where smaller droplets can form again. This process of wiping and formation of new small droplets exhibits low heat transfer resistance and is the reason for a large heat transfer coefficient.

Literature (Lee et al. 1998; Daniel et al. 2001; Liao et al. 2006; Zhu et al. 2009, Pratap et al. 2008) suggests various ways of controlling the motion of droplets. A simple approach for mobilization of drops is to incline the surface with respect to horizontal, wherein the gravitational body force is responsible for the droplet motion. Alternatively, one can introduce a variation of surface tension gradient on the substrate. Surface tension can be varied as follows. (1) Applying a large temperature gradient on the substrate, in which Marangoni effect leads to drop motion and (2) movement of micro-droplets underneath a horizontal surface by a variable-surface-energy coating, which creates a wettability gradient. To facilitate drop motion by artificially forming a wettability or surface energy gradient on the surface by suitable chemical treatment is a promising technique for drops motion as compared to applying a temperature gradient on the substrate. It is quite possible on copper and glass surfaces by depositing organic long chain monolayers (Subramanian et al. 2005; Pratap et al. 2008).

Daniel et al. (2001) and Bonner-III (2010) reported from experiments that the condensation on a wettability gradient surface is quite large as compared to a horizontal substrate without wettability gradient.

Against this background, dropwise condensation of water underneath a horizontal surface with unidirectional constant wettability gradient is numerical simulated by the mathematical model. Features of the condensation cycle underneath a horizontal substrate with a wettability gradient are shown in Fig. 3.16. These are similar to those of an inclined surface. The points of difference for a graded surface are: (1) drops shift towards the higher wettability side. Hence, drops of all the sizes are in motion, (2) velocity of drops depends on their size and position on the substrate, and (3) growth and sliding occur simultaneously.

The drop distribution, from initial nucleation to dynamic steady state, at selected instants of time, underneath horizontal substrates with constant wettability gradient, is shown in Figs. 3.17–3.19. The former shows the condensation pattern of water vapor at a saturation temperature of 303 K and degree of subcooling = 5 K underside of substrate having contact angles of 100° and 90°. The latter shows the condensation of water vapor at under similar conditions as in Fig. 3.17 but with contact angles of 110° and 100°. In both the surfaces, the wettability gradient is $0.33°\text{mm}^{-1}$. As drops grow, they become unstable and move towards the higher wettability side of the substrate. Therefore, on an average larger drops are present at the higher wettability side. The patterns of drops underneath a wettability gradient follow approximately the same trend as those underneath an inclined substrate.

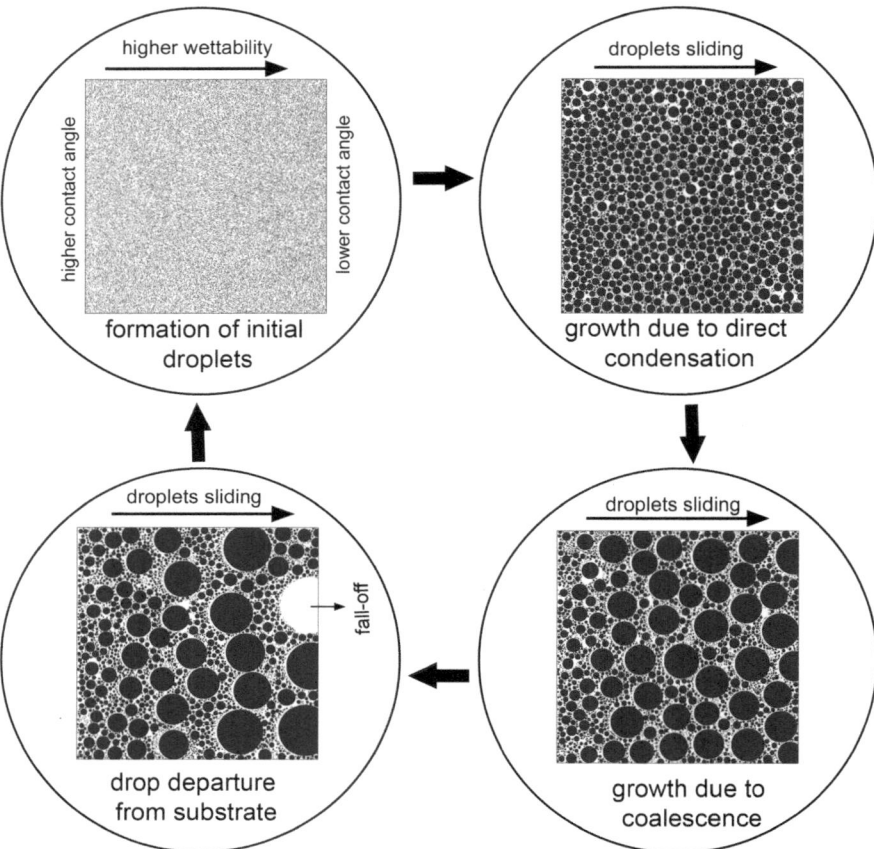

Fig. 3.16 Cycle of dropwise condensation observed in water vapor condensation underneath a horizontal substrate with unidirectional wettability gradient

The point of difference is that there is no critical size for drop instability. Every drop becomes unstable due to surface tension difference at the three-phase contact line. Gravity, viscosity, and surface tension forces are important to determine the terminal velocity of the drop. Hence, the sliding velocity depends on the size and position of the drop underneath the substrate. Fall-off is observed as a rule, on the higher wettability side of the substrate. Small drops present at the lower wettability side of the substrate are also highlighted in Figs. 3.17 and 3.18. On the higher wettability side, the driving force of drop becomes small, the drop cannot move, and it reaches criticality of fall-off (Fig. 3.17, 58 min). These results reveal that the micro-drop size can be moved as the hydrophobicity of wettability gradient substrate increases.

The comparison of dropwise condensation patterns of water vapor underneath horizontal and inclined nongraded substrates with a horizontal graded substrate is

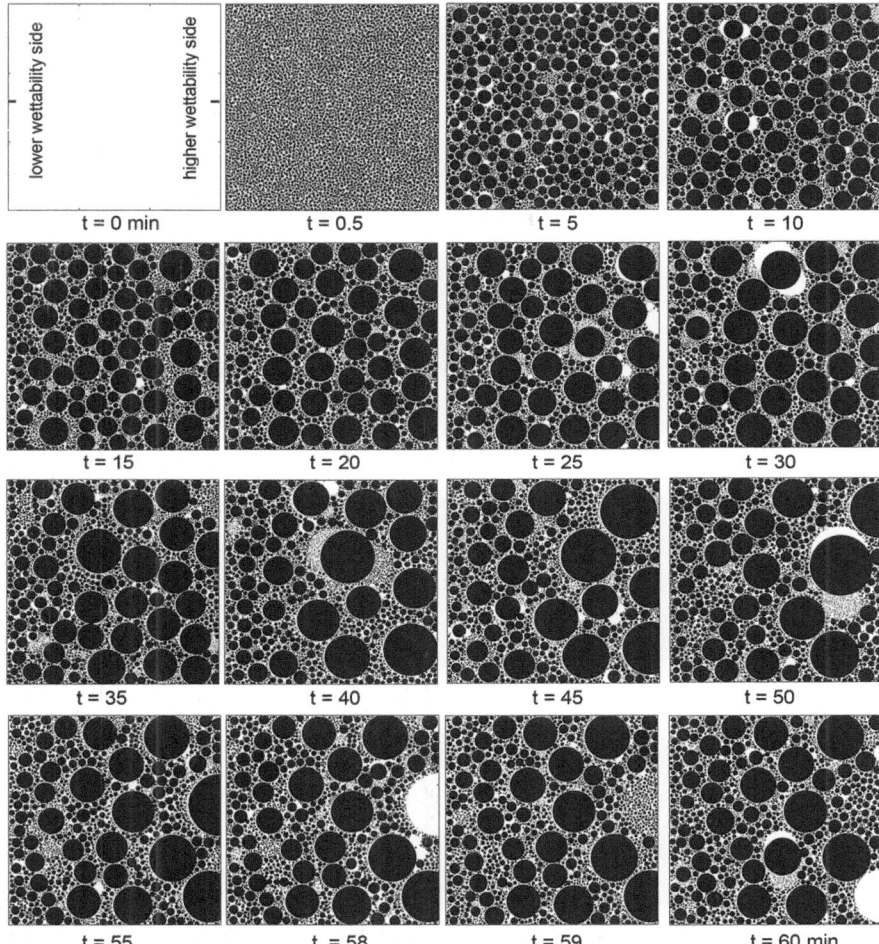

Fig. 3.17 Distribution in drops in dropwise condensation of water vapor underneath a horizontal substrate of wettability gradient $= 0.33°$/mm. Lower wettability side has a contact angle of $100°$, size of substrate is 30 mm $\times$ 30 mm, nucleation site density 10^6 cm^{-2} at saturation temperature 303 K and degree of subcooling 5 K

shown in Fig. 3.20. Spatial distribution of drops at an instant just before the first drop leaves the surface on a graded substrate, first slide-off from an inclined substrate and the first fall-off from a horizontal substrate, respectively, are compared in Fig. 3.20a.

As wettability gradient induces motion to the drops of every size, there exists an exposed virgin area behind every droplet on the graded substrate, as shown in Fig. 3.20a. Hence, the fraction of total area exposed for fresh condensation tends to be greater for a graded surface when compared to the other two configurations. The

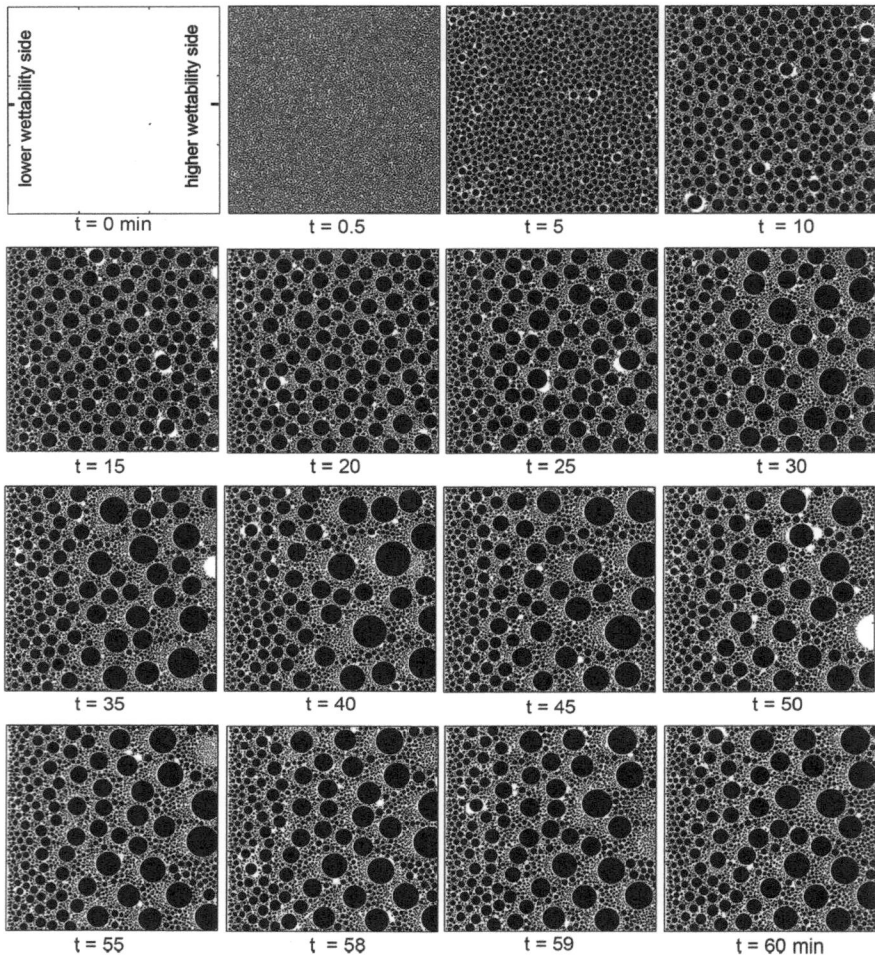

Fig. 3.18 Distribution in drops in dropwise condensation of water vapor underneath a horizontal substrate of wettability gradient $= 0.33°/\text{mm}$. Lower wettability side has a contact angle of $110°$, size of substrate is 30 mm $\times$ 30 mm, nucleation site density 10^6 cm^{-2} at saturation temperature 303 K and degree of subcooling 5 K

frequency of occurrence of a drop of a given radius, namely the histogram, on the three substrates, at an instant just before slide-off or fall-off criticality is attained, is shown in Fig. 3.20b. Drops slide-off for a graded surface as well as for the inclined. Drops fall-off from a horizontal surface. From Fig. 3.20b, it is clear that a graded substrate has a larger number of smaller sized drops as compared to the other two. Largest drops are formed on a horizontal substrate before they fall-off. This eventually leads to a slower condensation rate on the horizontal substrate; in this

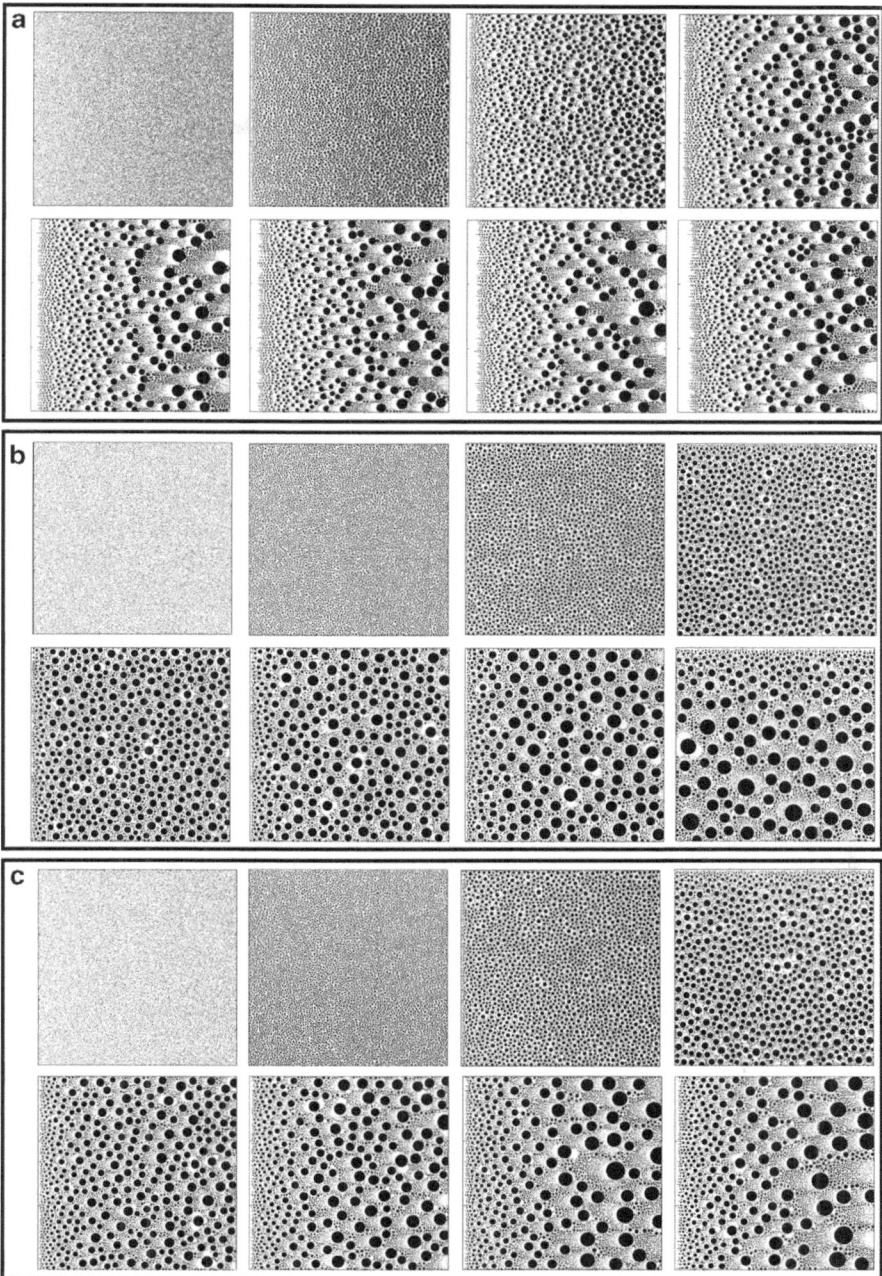

Fig. 3.19 Drop distribution condensed water vapor underneath a horizontal substrate with a wettability gradient. Contact angles are (**a**) 130° and 90°, (**b**) 130° and 100°, and (**c**) 130° and 120°

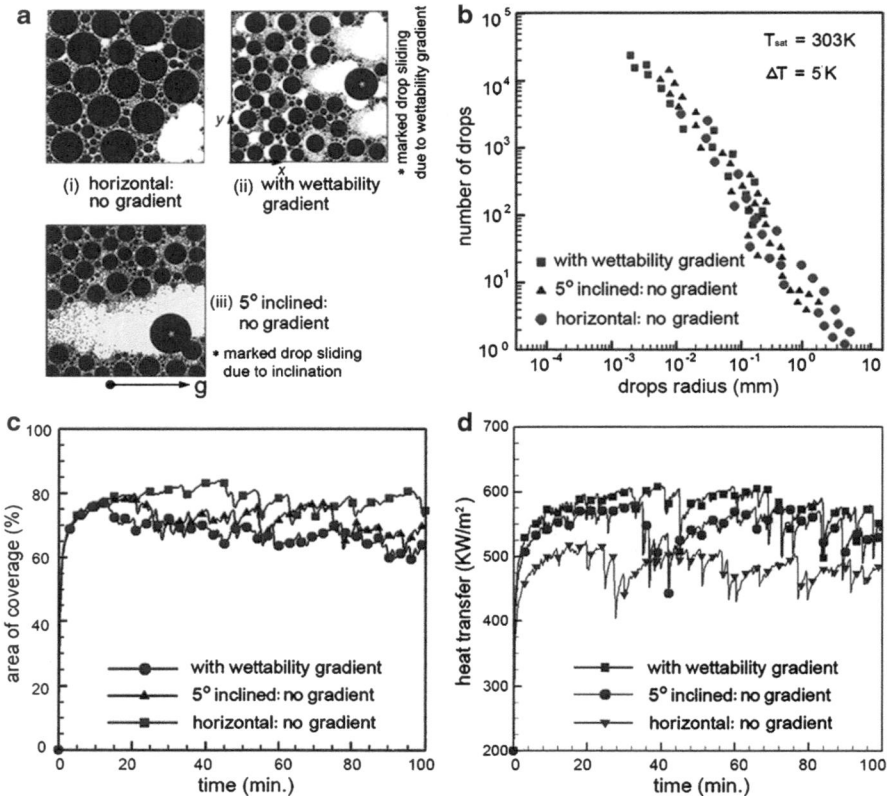

Fig. 3.20 Dropwise condensation of water vapor at 303 K with subcooling of $\Delta T = 5$ K underneath surfaces of various textures. (**a**) Drop distribution underneath various substrates soon after instability. (**b**) Number of drops as a function of drop radius, just before the largest drop leaves the surface. (**c**) Effect of the choice of the substrate on percentage area of coverage. (**d**) Variation of heat transfer coefficient with time over various substrates

regard the graded substrate shows a clear promise from a perspective of heat transfer enhancement.

The area of coverage created by the footprints of the drops on the substrate, as a function of time, is presented in Fig. 3.20c. As soon as the virgin substrate is exposed to vapor flux, direct condensation is initiated and the area coverage of drops increases rapidly. Later, coalescence dominates direct condensation, eventually leading to droplet criticality. The cycle is then established and the area coverage tends to stabilize. On a horizontal substrate, only a fall-off criticality is possible while on a graded substrate, a slide-off criticality is usually achieved first. During sliding motion, a droplet may fall-off in transit due to increase in its weight. A quasi-steady-state is eventually reached, after which the area coverage oscillates around an average value.

Results shown in Figs. 3.17, 3.18, 3.19, and 3.20 reveal that area coverage is smaller for the graded surface, making the exposed area greater than the other two surfaces considered. Further, drop instability in the form of a slide-off event is relatively early on the graded surface. As a direct consequence, heat transfer coefficient is expected to be higher for a surface with variable wettability. Heat transfer rates computed on these surfaces were found to be 450 (horizontal), 520 (inclined), and 540 (graded horizontal) in units of kW/m^2, with a subcooling of 5 K and a condensation temperature of 303 K.

3.2   Closure

Water vapor condensation underneath horizontal, inclined, and wettability gradient surfaces has been studied by numerical simulation. The effects of contact angle, contact angle hysteresis, inclination of the substrate, thermophysical properties of the working fluid, and saturation temperature of condensation are investigated. On the basis of numerical data, heat transfer coefficients of water vapor condensation are correlated. In order, the horizontal, inclined, and the graded surface experience (a) larger to smaller drop sizes, (b) longer to shorter cycle times, and (c) lower to higher heat transfer coefficients.

Chapter 4
Dropwise Condensation: Experiments

Keywords Textured surfaces • Condensation chamber • Imaging • Liquid crystal thermography • Validation of simulation with experiments

4.1 Introduction

It is possible to promote dropwise condensation satisfactorily under clean laboratory conditions. Despite sustained research over the past two decades, the prediction of the correct heat transfer rate during dropwise condensation over a surface remains a challenge (Tanasawa 1991; Stephan 1992; Rose 2002), mainly due to lack of knowledge of the local transport mechanisms of drop formation.

Experimental determination of the heat transfer coefficient during dropwise condensation is a difficult task because of the many intricacies involved in the process. The driving temperature difference is small, essentially resulting in a high heat transfer coefficient. Further, uncertainties associated with the microscale substructure of contact line shapes and motions, dynamic temperature variations below the condensing drops, effect of roughness and inhomogeneity of the substrate structure, control of true boundary conditions, microscale instrumentation, and transport dynamics of coalescence, merger, wipe-off, renucleation cycles, and the leaching rates of the promoter layer add to the difficulty in conducting repeatable experiments. Very high heat transfer rates (and therefore a very low temperature differential) coupled with the above factors also hinder generation of repeatable experimental data. Consequently, many conflicting experimental results have been published over the years, some results showing considerable scatter, Fig. 4.1.

Improved experimental techniques have led to reproducible and reliable experiment data to an extent; see, for instance, Le Fevre and Rose (1964, 1965) and Citakoglu and Rose (1968a, b). Several authors (Tanasawa 1991; Stephan 1992) reported that the heat transfer coefficient of dropwise condensation for steam at an atmospheric pressure, under the normal gravitational acceleration and on a vertical copper surface is about $230 \pm 30 \, \text{kW/m}^2 \, \text{K}$ in the heat flux range of 0.1–1 MW/m^2,

S. Khandekar and K. Muralidhar, *Dropwise Condensation on Inclined*   95
Textured Surfaces, SpringerBriefs in Applied Sciences and Technology 11,
DOI 10.1007/978-1-4614-8447-9_4, © Springer Science+Business Media New York 2014

Fig. 4.1 Experimental
results on dropwise
condensation (water,
~1 bar), as per Stephan
1992. *1* Hampson and
Özisik (1952), curve for two
different promoters;
2 Wenzel (1957); *3* Welch
and Westwater (1961);
4 Kast (1965); *5* Le Fevre
and Rose (1965); *6* Tanner
et al. (1968); *7* Griffith and
Lee (1967)

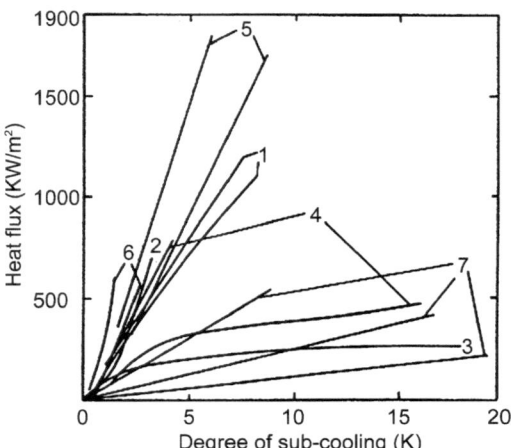

Fig. 4.2 Dropwise
condensation of steam on a
copper surface (*vertical*) at
atmospheric pressure
compared with the theory of
Le Fevre and Rose (1966),
adapted from Rose
et al. (1999)

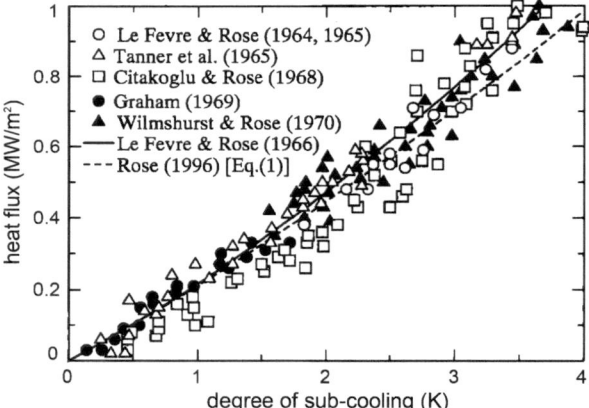

provided there is no effect of noncondensing gases and steam is approximately quiescent. Heat transfer coefficient of dropwise condensation of steam at atmospheric pressure has been summarized by Rose et al. (1999), Fig. 4.2. Several representative measurements shown in the figure are close to the theory Le Fevre and Rose (1966). Data for other vapors and measurement under other thermal conditions are still scarce.

With the advent of newer coating/manufacturing and nanoscale fabrication techniques, promoting long-term sustainability of dropwise condensation by chemical coating holds considerable prospect for enhancing heat transfer. An example of enhanced performance of compact steam condensers having chemically coated flow passages of only a few millimeters width is demonstrated by Majumdar and Mezic (1999). It is necessary to understand the effect of various parameters on heat transfer during dropwise condensation as reviewed next.

Marto et al. (1986) tested several polymer gold and silver coatings for sustaining dropwise condensation of steam and reported that the heat transfer coefficient in dropwise condensation is as high as six times when compared to the filmwise. Zhao et al. (1996) reported heat transfer coefficient on Langmuir-Blodgett treated surface to be than 30 times more than that of filmwise condensation on a bare surface. Koch et al. (1998) showed the effect of hydrophobicity on heat transfer coefficient on a chemically textured vertical substrate. Heat transfer coefficient was found to decrease with an increase in wettability.

Ma and Wang (1999) proposed that the heat transfer coefficient increases with the increase in the surface free energy difference between the condensate and the condensing substrate. Hence, surface modifications for promoting dropwise condensation by silanation and ion implantation are of particular interest. These would yield continuous dropwise condensation along with a high heat transfer coefficient. Leipertz and Cho (2000) reported heat transfer rates on several metallic substrates (copper, titanium, aluminum, high-grade steel, and hastelloy) treated by ion implantation. Ions considered were nitrogen, oxygen, and carbon ions, with varying ion density. Das et al. (2000) applied an organic self-assembled monolayer coating to enhance the dropwise condensation, the corresponding increase in the heat transfer coefficient being a factor of 4.

Vemuri et al. (2006) performed a condensing experiment over various coated substrates and reported long-term sustainability and enhancement of heat transfer coefficient. The authors coated a copper substrate with self-assembled mono-layers (SAMs) of n-octadecyl mercaptan and stearic acid. An increase in heat transfer coefficient by a factor of 3 was reported as compared to a bare copper substrate. Ma et al. (2008) experimentally studied dropwise condensation on a vertical plate for a variety of noncondensable gas (NCG) concentration, saturation pressure, and surface subcooling. A fluorocarbon coating was applied to promote dropwise condensation. Departure of drops was inferred as the dominant factor for the steam–air condensation heat transfer enhancement.

Rausch et al. (2007; 2010a, b) observed that the heat transfer coefficient on an ion-implantation surface is more than five times than that of filmwise condensation. Ion implanted metallic substrates have stable condensation as well as high heat transfer coefficient over a long time duration. Chen et al. (2009) experimentally investigated the effects of various chemical coatings and their long-term durability on the dropwise mode of condensation. A reduction in the heat transfer coefficient was seen with the elapsed condensation time, suggesting possible leaching of the chemical coating. Dietz et al. (2010) investigated droplet departure frequency using electron microscopy to understand enhancement of dropwise condensation on superhydrophobic surfaces. A reduction in drop departure size shifts the drop size distribution to smaller radii, which may enhance the heat transfer rate.

Ma et al. (2012) investigated experimentally the heat transfer characteristics in the presence of a noncondensable gas (NCG) on superhydrophobic and hydrophobic surfaces including the wetting mode evolution on the roughness induced superhydrophobic surface. Superhydrophobic surfaces with high contact angle ($>150°$) and low contact angle hysteresis ($<5°$) were seen to be an ideal condensing surface to promote dropwise condensation of water and enhance heat transfer. With increasing NCG concentration, the droplet undergoes transition from the Wenzel to Cassie-Baxter mode.

Fig. 4.3 Dependence of heat transfer coefficient on pressure for steam condensation on copper substrate in the form of drops (Rose et al. 1999)

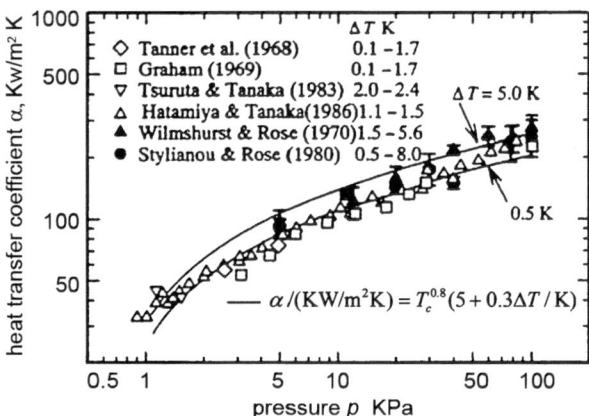

Miljkovic et al. (2012) studied the effect of droplet morphology on heat transfer during dropwise condensation on superhydrophobic nanostructured surfaces. These surfaces were designed to be Cassie stable and favored the formation of suspended droplets on the top of the nanostructures as compared to the partially wetting droplets which locally wetted the base. Cassie stable droplets were seen to have minimal contact line pinning and promoted passive droplet shedding at sizes smaller than the characteristic capillary length. However, the gas films underneath such droplets significantly hindered the overall heat and mass transfer performance.

4.1.1 Thermophysical Properties of Condensate

Several sets of results are available for dropwise condensation of steam on copper at atmospheric pressure conditions. Fewer results are available on the heat transfer coefficient at pressures lower than one atmosphere (Tanner et al. 1968; Graham 1969; Wilmshurst and Rose 1970; Tsuruta 1993; Hatamiya and Tanaka 1986). They show a tendency of decreasing heat transfer coefficient with decreasing pressure, Fig. 4.3.

Condensation of other vapors in the form of drops has been reported by several authors. Wilmshurst and Rose (1974) performed condensation experiments of aniline and nitrobenzene on PTFE coated substrate. Stylianou and Rose (1983) reported condensation of ethylene-glycol on a copper substrate. Utaka et al. (1987, 1994) performed a condensation experiment with propylene-glycol, ethylene-glycol, and glycerol vapors on copper substrate using a monolayer type promoter below atmospheric pressure. Quantitatively, the heat transfer characteristics for organic vapors differ from those for steam due to wide variation of physical properties. Owing to the lower liquid thermal conductivity of the organic fluids, relatively low heat transfer coefficients are to be seen in comparison with steam. For moderate subcooling, the heat transfer coefficient for dropwise condensation is significantly larger than for film condensation. It can also be seen that the surface subcooling ranges of ideal dropwise condensation differ widely, depending on the choice of the fluid. For a fluid of higher surface tension such as glycerol, dropwise condensation is maintained for larger surface subcooling, compared to a lower surface tension liquid such as propylene-glycol.

Dropwise condensations of low Prandtl number vapors are scarcely presented in the literature. In many situations, a singular behavior is observed for low Prandtl number systems, for example, liquid metals where $Pr \approx 0.01$. Moreover, condensation of liquid metals also plays an important role in many engineering processes. Only a few researchers have considered dropwise condensation of metal vapors. Bakulin et al. (1967) reported the effect of a noncondensable phase on temperature drop at the liquid–vapor interface during the dropwise condensation of sodium, potassium, and lithium. Interfacial resistance to mass transfer at the liquid–vapor interface was seen to play an important role in the condensation of metals vapors. Rose (1972) modified the previously reported theory of dropwise condensation and showed that the degree of subcooling affects heat transfer in mercury, though the effect is less than in water. Necmi and Rose (1977) experimentally measured vapor-to-condensing surface temperature difference and the corresponding heat flux for various vapor pressures during dropwise condensation of mercury on a vertical substrate. Niknejad and Rose (1984) compared the experimental data of mercury with their own theory of dropwise condensation developed for water and found significant differences. Literature on dropwise condensation of other metal-vapors such as sodium, potassium, and bismuth is not available, though the liquid phases of these substances have rather large surface tension.

Many researchers (Takeyama and Shimizu 1974; Tanasawa and Utaka 1983; Tanasawa 1991; Rose 2002, 2004; Lan et al. 2009) reported increase in heat flux along with an increase in the surface subcooling. At a higher subcooling, the rate of drop nucleation may increase because the minimum radius of drop must be smaller.

4.1.2 Physicochemical Properties of Substrate

The substrate hydrophobicity, contact angle hysteresis, and state of drop (Cassie versus Wenzel) on or underneath a substrate depend on its physicochemical properties with respect to the condensing fluid. These parameters play an important role in dropwise condensation. Many researchers (Lee et al. 1998; Lara and Holtzapple 2011; Baojin et al. 2011) reported that high contact angle ($>150°$) and low contact angle hysteresis ($<5°$) is an ideal combination for a condensing surface. Neumann et al. (1978) reported that heat-transfer during dropwise condensation of water vapor strongly depends on contact angle hysteresis. This is because the surface conductance increases with decreasing contact angle hysteresis. Kim and Kim (2011) reported a strong effect of contact angle on the heat transfer rate. A large contact angle leads to the enhancement of heat transfer. Miljkovic et al. (2012) reported that the heat transfer coefficient of dropwise condensation depends on the morphology of droplets on the substrate. The initial growth rates of partially wetting droplets (Wenzel) were six times larger than the suspended droplets (Cassie). Experimental results showed that partially wetting droplets (Wenzel) had 4–6 times higher heat transfer rates than the suspended droplets (Cassie).

Although the heat transfer coefficient of dropwise condensation of vapors are strong functions of physicochemical properties such as contact angle and contact angle hysteresis, there is no correlation linking the heat transfer coefficient with contact angle, contact angle hysteresis, and droplet morphology on the substrate. Such a relationship has been examined in the present monograph.

4.1.3 Substrate Having a Wettability Gradient

In dropwise condensation, liquid droplets forming on a subcooled nonwetting surface are removed from the surface by gravitational forces when the droplets reach a critical mass. The dependence on gravity for liquid removal limits the utilization of dropwise condensation in low gravity aerospace applications and horizontal surfaces. Various authors (Zhao and Beysens 1995; Daniel et al. 2001; Liao et al. 2006) have applied a novel passive technique based on surface energy gradient in the condensing surface to remove droplets. Daniel et al. (2001) observed the random movements of droplets to be biased towards the more wettable side of the surface. Powered by the energies of coalescence and directed by the forces of the chemical gradient, small drops (0.1–0.3 mm) display speeds that are faster than those of typical Marangoni flows. Wettability gradient on a horizontal substrate has implications for passively enhancing heat transfer in heat exchangers and heat pipes. Bonner (2009) verified experimentally that a wettability gradient substrate has high heat transfer compared to a horizontal substrate. Gu et al. (2005) and Bonner-III (2010) enhanced heat transfer of a condensing system by creating energy gradient on the condensing substrate.

4.1.4 Substrate Orientation

The study of orientation of the cold substrate is important in dropwise condensation and enhancement of heat transfer. Many researchers (Citakoglu and Rose 1968b; Izumi et al. 2004; Leipertz and Fröba 2006) have reported high rate of water vapor condensation on vertical substrates for a given degree of subcooling. Leipertz and Fröba (2008) reported the following correlation for the heat transfer coefficient in dropwise condensation as a function of the inclination of the substrate:

$$h_c(\alpha) = h_c(90°) \cdot [\sin \alpha]^\kappa \qquad (4.1)$$

Fig. 4.4 Variation of heat
transfer coefficient with
respect to angle of
inclination of the substrate
(Leipertz and Fröba 2008)

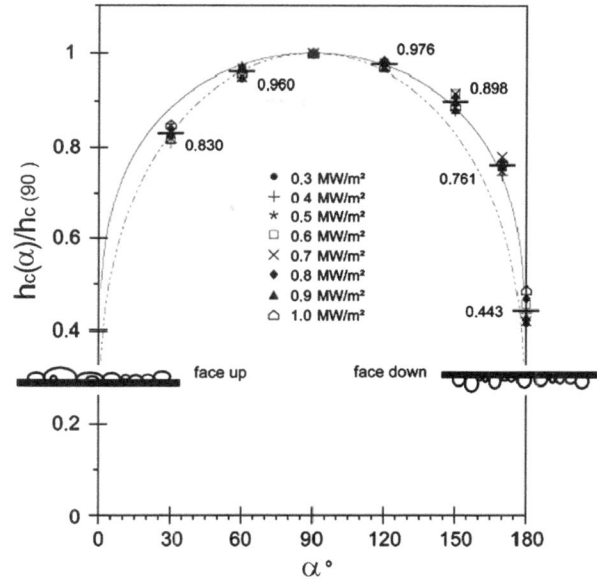

Here, $h_c(\alpha)$ is the heat transfer coefficient of dropwise condensation at angle α
and $h_c(90°)$ corresponds to that of a vertical substrate.

In Fig. 4.4, the value of κ was ~0.270 for the dashed line and ~0.176 for the solid
line. The angle of inclination is defined to be 0° for the horizontal surface with the
drops on the upper side of the substrate (sessile mode) and 90° for a vertical. From
90° onwards, drops form on the lower side of the substrate and the 180° horizontal
substrate refers to the pendant mode of dropwise condensation. The heat transfer
data for water vapor in dropwise condensation with respect to orientation (sessile
and pendant) are shown in Fig. 4.4. It is clear that the pendant mode over a
horizontal substrate yields a higher heat transfer coefficient as compared to the
sessile. Heat transfer coefficient is the highest for a vertical substrate and decreases
with increasing inclination. For an inclined substrate, the surface is swept clean of
drops and this renewal of the growth process is responsible for a higher heat transfer
coefficient. In contrast, drops over a horizontal surface become large and fall-off by
gravity in the pendant mode or spread over the substrate and cover it by a layer of
the condensate liquid, in case of sessile drops. In both the cases, the surface is not
regularly refreshed by fresh condensation, resulting in a lower heat transfer coefficient. Tanasawa et al. (1976) measured the dependence of heat transfer coefficient
on the departing drop diameter. Authors reported that the heat transfer coefficient is
proportional to the departing drop diameter to the power of about ~0.3. Lawal and
Brown (1982) and Briscoe and Galvin (1991b) reported that a pendant drop is less
stable as compared to a sessile drop on an inclined substrate, suggesting that heat

transfer during dropwise condensation underneath an inclined substrate is marginally better than its counterpart above the surface. Therefore, surface orientation is an important parameter in the enhancement of heat transfer coefficient in dropwise condensation. These issues are addressed in the present study.

4.2 Surface Preparation Techniques

In process equipment, dropwise condensation can be realized by suitably treating the condensing surface. The treatment will ensure partial wetting of the surface by the condensate liquid in the sense that the contact angle greater than 90° is achievable. The wetting characteristics of condensate over the cold substrate can be broadly controlled by two different means:

1. Modify the surface of the substrate, or
2. Alter the condensing vapor chemically; say by injecting a chemical which promotes nonwetting behavior.

Other methods that rely on changing the pH value of the condensate can be used so that dropwise condensation can be controlled by using an electrical potential (electro-wetting), changing the condensing temperature, and other techniques. Among these methods, substrate modification has emerged as the most popular and effective strategy.

A good drop promoter technique should be long lasting, involving low surface energy, low contact angle hysteresis, and low thermal resistance. The method should be easy to apply, nontoxic, and must be compatible with the system in which it is used, i.e., it should not impair the proper functioning of the other parts of the system. Superhydrophobicity appears ideal to promote continued dropwise condensation which results in rapid removal of condensate drops; however, such promotion has not been reported on engineered surfaces. For any technique used for promoting dropwise condensation, the longevity of the textured surface is critical. With the advent of newer manufacturing, coating, and nanoscale fabrication techniques, surface treatment of the substrate holds considerable prospect in terms of providing the required long-term sustainability of dropwise condensation.

There have been two generic methods that can be used to modify the wettability of the substrate. The first one is chemical grafting or adsorbing molecules with wetting characteristics of their own (chemical texturing). The second is to texture the surface by altering the surface topography/roughness by patterning, called physical texturing. Roughening a surface will increase wettability, in general, unless special patterns of the right scale are employed. In contrast, chemical coatings have gained prominence because of the larger choices available and are reviewed here.

4.2.1   Chemical Texturing

Chemical texturing can be created by coatings, such as organic compounds, with hydrophobic groups (Blackman et al. 1957; Watson et al. 1962; Ma and Wang 1999; Vemuri et al. 2006), inorganic compounds (Erb 1965; Erb and Thelen 1965, 1966; Zhao et al. 1996), polymers (Marto et al. 1986; Zhang et al. 1986; Mori et al. 1991; Ma et al. 2002) or special surface alloys (Erb 1973; Zhao et al. 1991; Koch et al. 1997; Ma et al. 2000a, b). Other coating materials include, for example, Teflon (Stylianou and Rose 1980). These surfaces are created by preparing a weak solution of Teflon (AF1600) in FC-75. The samples are dip-coated in this solution with different pulling speeds to achieve the desired film thickness. After dip coating, the samples are annealed in a furnace for ~10–30 min at temperatures ranging from 100 to 300 °C (Ma and Wang 1999; Ma et al. 2000b).

Though simple in concept, such surfaces suffer from long time sustainability issues that do not allow application to real-life, large-scale processes. Leaching by the motion of drops over the surface can also result in degradation of the coating. Amorphous hydrogenated carbon films (a-C:H) with diamond-like mechanical properties have been modified by adding new elements to the film, e.g., silicon or fluorine (Grischke, et al. 1994), reducing its surface energy. These coatings have been studied for their heat transfer characteristics. Such coatings are mechanically and chemically stable but introduce an additional thermal resistance. This drawback can be overcome by other surface modifications which do not form an additional layer. Ion-implantation is an example that has been tested successfully by Zhao et al. (1990), Zhao et al. (1991), Leipertz and Cho (2000), and Zhao and Burnside (1994).

In the present work, a hydrophobic surface is prepared by coating it with a self-assembled monolayer (SAM). The condensing pattern obtained underneath this chemically coated substrate is used for validation of simulation data. Among the SAMs, Octa-decyl-tri-chloro-silane (OTS) was found to yield the best quality surface for dropwise condensation because of the smallest contact angle hysteresis.

The preparation of the hydrophobic substrate by self-assembled monolayers uses the chemical vapor deposition process, Fig. 4.5a. The samples were first cleaned by sonicating them in ethanol, acetone, and toluene bath for 3–10 min respectively. The substrate was dried with compressed nitrogen gas carefully while changing from one solvent to another. Subsequently, the samples were cleaned by an oxygen plasma torch followed by a dry CO_2 snow-jet (Cras et al. 1999). The samples were kept in Piranha solution (50 % H_2O_2 and 50 % H_2SO_4 by volume) for 2–4 h.

Piranha solution is highly oxidizing and requires special care in handling. Surface energy modification was accomplished by coating samples with self-assembled monolayers (SAM). SAMs are formed spontaneously by chemisorption and self-organization of functionalized and long-chain organic molecules on an appropriate surface. Octadecyl-tri-chloro-silane (OTS), di-methyl-chloro-silane (HMS), trichloro-silane (MTS), and propyl-tri-chloro-silane (HTS) were used as

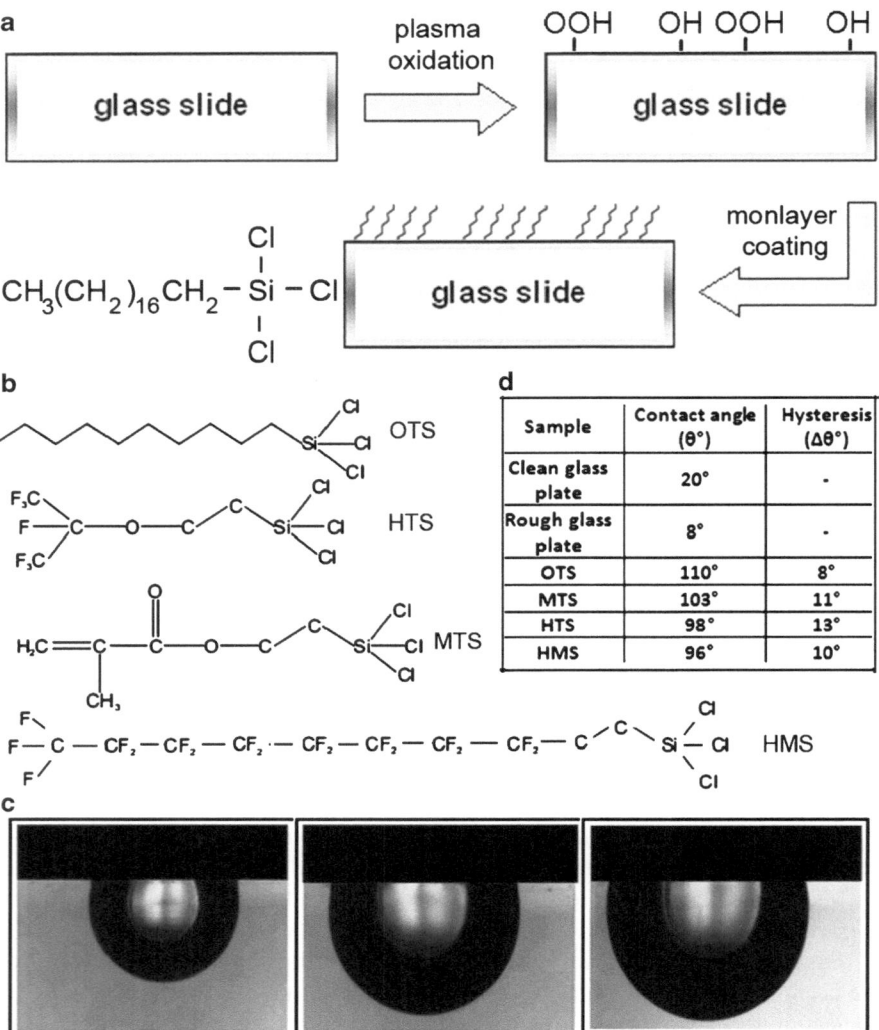

Fig. 4.5 (**a**) Schematic diagram explaining the chemical vapor deposition process. (**b**) Representations of various self-assembled monolayer (silane) molecules. (**c**) Image of pendant drop of volume 5 μl, 10 μl, and 15 μl respectively on HMS textured substrate. (**d**) Measurement data

SAMs, Fig. 4.5b. To deposit OTS on a surface, the cleaned substrates were kept in a solution of 60 ml bi-cyclo-hexane, 35 drops carbon-tetrachloride, and 20 drops OTS. During this time, the OTS molecules bond covalently on silicon dioxide substrates. After a few minutes, the substrates were taken out of the silane solution and rinsed with chloroform.

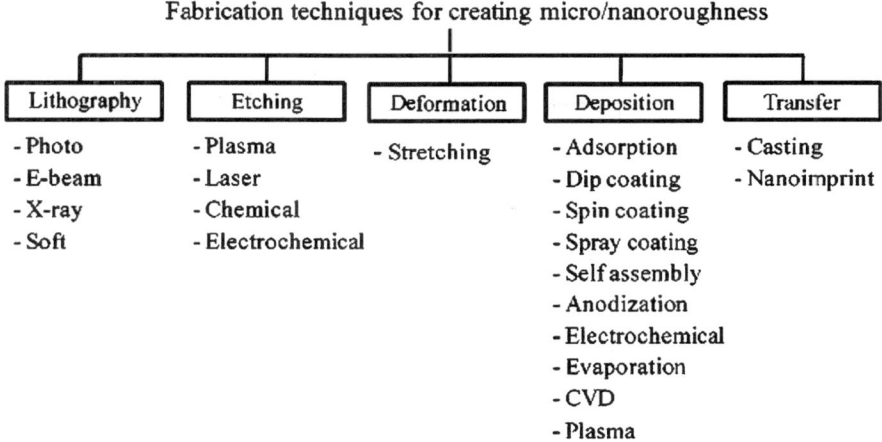

Fig. 4.6 Typical methods to fabricate micro/nanoroughened surfaces (Bhushan and Jung 2011)

To coat the surface with di-methyl-chloro-silane (HMS), trichloro-silane (MTS), and propyl-tri-chloro-silane (HTS), the cleaned samples were kept in a desiccator together with a small quantity of the desired silane. Silane vaporizes in the closed environment of the desiccator and gets deposited on the substrate. After 15–20 min of evaporation, a silane monolayer gets bonded covalently with the oxide surface (Genzer and Efimenko 2000). After taking out the samples from the desiccators, they are rinsed by chloroform. Co-evaporation of various silanes can also be carried out to achieve intermediate surface energy but at the cost of a higher contact angle hysteresis. In Fig. 4.5c, d, the image of a pendant drop for various chemically textured substrates, along with the measurement data for apparent contact angle, is shown. For a good hydrophobic coating, contact angle hysteresis should be as small as possible.

4.2.2 Physical Texturing

Fabrication of hydrophobic surfaces by physical texturing is, in principle, quite simple. It can be generated by creating a suitable roughness. A review of the subject (Nakajima et al. 2001) reveals a wide range of methods for producing physical texturing/topography roughness on common enginenring materials. Some common methods available to create roughness distribution on a substrate are shown in Fig. 4.6. The pros and cons of these available techniques are summarized in Table 4.1.

Most researchers (Sommers and Jacobi 2006, 2008) reported hydrophobic and superhydrophobic surfaces produced by etching and lithography. Some (Lau et al. 2003; Chen et al. 2007; Hsieh et al. 2008; Boreyko and Chen 2009) reported

Table 4.1 Pros and cons of various surface fabrication techniques

Techniques	Pros	Cons
Lithography	Accuracy	Large area, slow process, high cost
Etching	Fast	Contamination, less control
Deposition	Flexibility, cheap	High temperature, less control
Self-assembly	Flexibility, cheap	Require suitable precursor

continuous dropwise condensation on a superhydrophobic surface with short carbon nanotubes deposited on micromachined posts, a two-tier texture mimicking lotus leaves. Surface preparation by physical texturing as reported in the literature is reviewed below.

Jessensky et al. (2003) introduced a new technique for the fabrication of a superhydrophobic surface by anodic oxidation of metals such as aluminum, titanium, tungsten, and hafnium. These metals may all be anodically oxidized when put into an electrolyte. Such surface treatments are common in the industry, as anodization of aluminum and titanium creates hard scratch-resistance and protects the underlying surface from further oxidation. When refined to a specific processing regime, anodization results in the formation of a highly ordered nanostructure.

Miwa et al. (2000) prepared various superhydrophobic films of different surface roughness. The relationships between the sliding angle, the contact angle, and the surface structures were investigated. Sliding angle of water droplet was seen to decrease with increasing contact angle. Micro structures revealed that the surface texture traps air and assists in the preparation of low-sliding-angle surfaces.

Sommers and Jacobi (2006), describe photolithographic techniques to obtain micropatterns on aluminum surfaces with parallel grooves, 30 μm wide and tens of microns in depth. Experimental data show that a droplet placed on the microgrooved aluminum surface using a microsyringe exhibits an increased apparent contact angle. For droplets condensed on these etched surfaces, more than a 50 % reduction in the volume needed for the onset of sliding is obtained.

Liu et al. (2006) prepared micro–nanoscale binary structured composite particles of silica/fluoropolymer by using an emulsion-mediated sol–gel process. It is composited on various substrates by using simple spray coating or spin coating methods to create superhydrophobic surfaces. Results show that the static contact angle of water on the substrate is larger than 150°.

Boreyko and Chen (2009) generated a superhydrophobic substrate, composed of two-tier roughness with carbon nanotubes deposited on silicon micropillars and coated with hexedecanethiol. Continuous dropwise condensation was spontaneously formed on a superhydrophobic surface without any external forces. Spontaneous drop removal resulting from the surface energy released upon drop coalescence led to an out-of-plane jumping motion of the coalesced drops at a speed as high as 1 m/s.

Dietz et al. (2010) reported a novel technology to achieve superhydrophobic coating with microscopic roughness on a copper surface. A layer of verdigris was grown on fresh pure copper surface by exposing copper to air and a mist of acetic

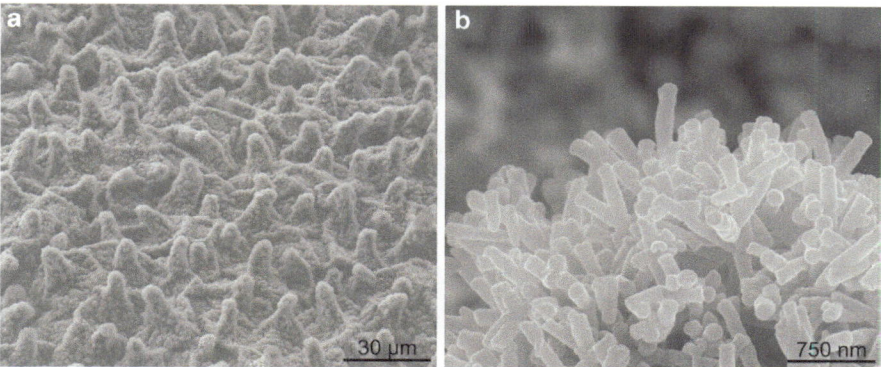

Fig. 4.7 SEM images of a lotus leaf, *Nelumbo nucifera*. (**a**) The surface is covered with hierarchical roughness so that the micro bumps and the basal area are entirely decorated with nanoprotrusions. (**b**) Randomly oriented nanocylinders that cover a micro bump, Cha et al. (2010)

acid solution. The coating was oxidized to CuO. A self-assembled monolayer coating of n-octadecanethiol was obtained on the outermost surface. Results showed that the static contact angle of a water drop was 153.1° ± 1.7°.

Cha et al. (2010) fabricated six different surfaces, one natural and five artificial. As a natural hydrophobic surface, a lotus leaf, *Nelumbo nucifera* was used. A lotus leaf collected from a local pond was cleaned with an air gun to remove dust particles. The leaf surface was covered with hydrophobic epicuticular wax crystals and its water repellency was enhanced by the intrinsic surface structure. A scanning electron microscope of a lotus leaf surface is shown in Fig. 4.7.

Artificial surfaces were prepared bare with silicon wafers, single-roughness surfaces with micropillar arrays, single-roughness surfaces with nanoscale pillars, hierarchical surfaces with micropillars decorated with nanoprotrusions only on their tops (surfaces with partial dual roughness), and hierarchical surfaces with nanoscale roughness on both micropillar tops and bases (surfaces entirely with dual roughness). The process is schematically shown in Fig. 4.8. Contact angle of water drops on these surfaces varied from 140° to 170°, depending on the micro pillar density. On micro-/nanostructured surfaces, the condensate drops prefer the Cassie state which is thermodynamically more stable than the Wenzel state.

McCarthy et al. (2010) reported fabrication and characterization of biomimetic superhydrophobic surfaces synthesized using self-assembly and metallization of the Tobacco Mosaic Virus (TMV) onto micro pillar arrays. Superhydrophobic surfaces with static contact angles greater than 150°, and droplet hysteresis less than 10°, were seen to resist wetting and exhibit self-cleaning properties.

Despite breakthrough in nanomachining, there is no literature that has reported generation of a hydrophobic surface by conventional machining process. Bhutani et al. (2013) used hand lapping process on aluminum and copper substrate to make the surface hydrophobic. Lapping pastes of three different grades were used to produce surface roughness of the order of 0.5, 1.5, and 3.5 μm (RMS). The highest contact angle obtained was 95° on these substrates.

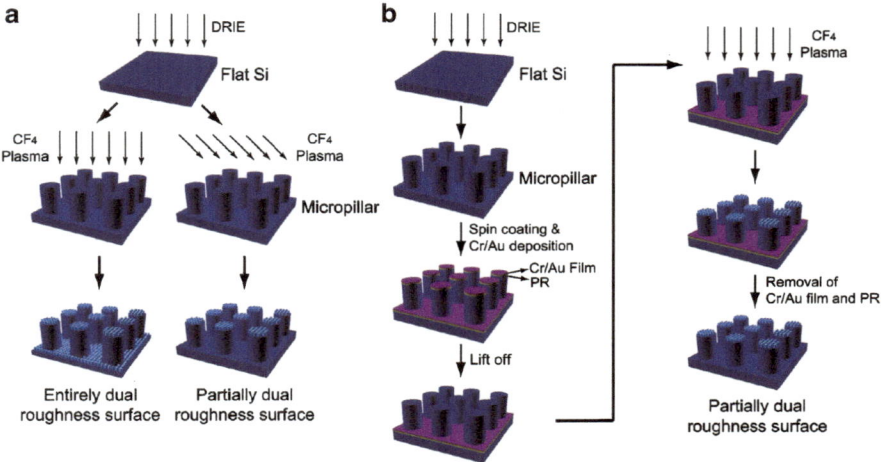

Fig. 4.8 Fabrication process of dual-roughness surfaces. (**a**) Fabrication of the surface with dual roughness with the direct incidence of CF₄ plasma. (**b**) Fabrication of the surface with partially dual roughness via masking the basal area with a Cr/Au layer, Cha et al. (2010)

4.3   Experiments on Chemically Textured Surfaces

The experimental apparatus was designed to study dropwise condensation under controlled conditions on the underside of a cold substrate, and is schematically shown in Fig. 4.9. The setup primarily consisted of the main cylindrical stainless steel vacuum chamber (better than 10^{-5} mbar abs.) of inner diameter 180 mm and length 120 mm (Fig. 4.9a, b). It was closed from the two ends by specially designed flanges. The lower flange was fitted with a $\lambda/4$ optical viewing window. Typical photographs of condensing droplets are shown in Fig. 4.9c. In addition, the optical window also had an annular space around, wherein the working fluid inventory (distilled and deionized water) was stored. A circular, 1.5-mm thick mica strip heater (OD $= 70$ mm, ID $= 40$ mm) was attached outside the annular space to give the necessary heat input, as shown in the cut section of the experimental setup in Fig. 4.9d. The upper end of the main vacuum chamber was closed with a polycarbonate square flange with an inbuilt cavity wherein cold water was circulated to maintain constant temperature boundary conditions. The condenser capacity was at least 20 times that of the maximum expected heat transfer rate. The chemically coated glass substrate of 100 mm $\times$ 100 mm was integrated on the upper flange, as shown in Fig. 4.9d. Connections for evacuation, pressure transducer, and temperature sensors were provided on the main condensing chamber wall. The temperature of the condensing vapor was measured with one K-type thermocouple (Omega®, 0.5 mm diameter) of accuracy $\pm$ 0.2 °C after calibration. It was placed centrally in the chamber at a distance of 25 mm from its side wall. The condensing chamber

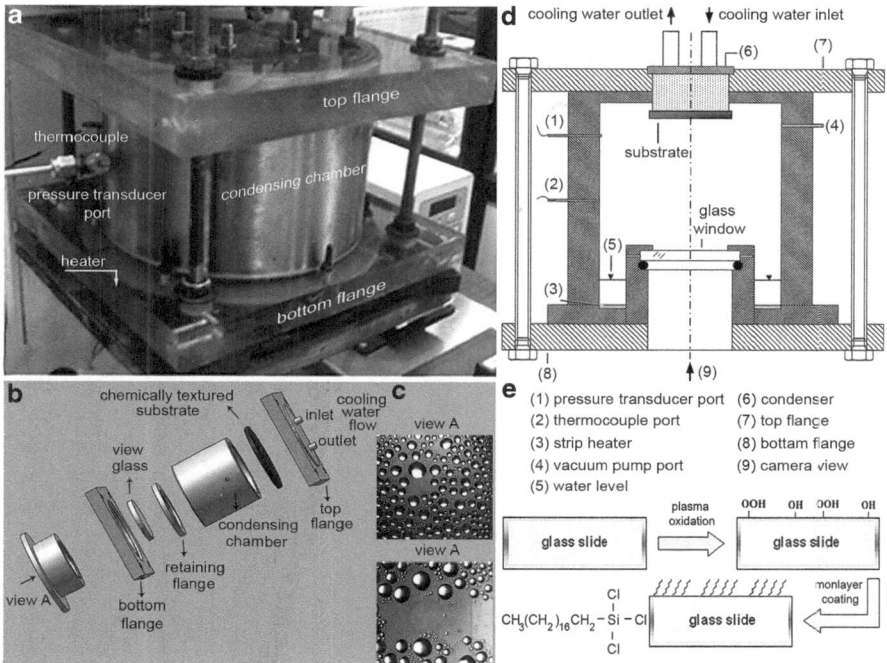

Fig. 4.9 Details of the experimental setup to study dropwise condensation under controlled conditions underneath a substrate. (**a**) Photograph shows the details of the main condensing chamber. (**b**) Exploded view of the condensing chamber. (**c**) Typical images of the condensing droplets at two different times, as captured from View-A. (**d**) Cross sectional view of condensing chamber. (**e**) Schematic diagram explaining the CVD process (Sikarwar et al. 2011)

pressure was measured by an absolute pressure transducer (Honeywell, accuracy 0.1 % FS, NIST traceable calibration, range 0–1.2 bar). Online data acquisition was carried out with 16-bit PCI-4351 card (National Instruments®). The entire assembly could be tilted to any desired inclination between 0° and 50°. A color CCD video Camera (Basler® A202KC with 1,024 × 1,024 pixels at 100 fps) was used to capture the images of the drops forming on the underside of the chemically textured substrate (View A, Fig. 4.9b). Length scales were calibrated by imaging a grid with known periodicity. Diffused white light source symmetrically placed around the camera was directed on the substrate from the optical window on the bottom flange so as to maintain a near parallel and symmetric beam on the droplets ensuring proper contrast level for subsequent edge detection.

The substrate preparation involved coating the glass surface using chemical vapor deposition of silane molecules. The chemical vapor deposition setup consisted of a vacuum pump (rotary vane rougher pump coupled with diffusion pump, ultimate vacuum level ~10^{-5} mbar), plasma oxidizer (with RF generator of 6–18 W power, frequency 8–12 MHz), and a desiccator. Inside the reactor, which was maintained at low vacuum pressure, the high frequency oscillating electromagnetic field ionized the silane molecules forming plasma. This interacted with the glass substrate by removing organic contamination from its surface. The high energy plasma particles combined with the contaminants to form carbon dioxide (CO_2) or methane (CH_4). The physicochemical characteristics of surfaces were modified by adsorption or chemisorption as follows:

The silanation process is explained schematically in Fig. 4.9e. Octyl-decyl-trichloro-silane ($C_{18}H_{37}C_{13}Si$ supplied by Sigma Aldrich®) was used as the coating material on the glass substrate. Before keeping the substrate for 30 min inside the CVD reactor, the substrate was cleaned by dipping it in a pirani solution (sulphuric acid and hydrogen peroxide in the ratio 3:1 by volume) for 2 h, thereafter washing it with distilled water and drying it in nitrogen. Nascent oxygen released when sulphuric acid reacted with hydrogen peroxide cleaned the surface. The silane molecules attached themselves to the plasma cleaned glass plate, which was kept inside the CVD chamber, by a self-assembled monolayer process. After preparing the surface, the static, advancing, and receding contact angles, respectively, of a pendant water drop for horizontal and inclined substrates were measured by a goniometer device that had a special attachment for inclining the substrate with respect to horizontal.

4.3.1 Experimental Methodology

Dropwise condensation of distilled and deionized water, underneath a horizontal substrate and an inclined substrate having various inclination (10°, 15°, and 30°), was carried out underneath a glass substrate which was coated with Octyl-decyl-trichloro-silane ($C_{18}H_{37}C_{13}Si$). The chamber temperature was maintained at 27 °C in all the experiments with cold substrate maintained at 22 °C. The static contact angle of water drop placed on the chemically textured substrate was measured to be 96° ± 0.5° for droplet volume range of 50–100 µl. Dropwise condensation was achieved at the desired saturation pressure by controlling the coolant temperature and the heat throughput. Once quasi-steady state was reached, the correspondence between the saturation pressure and the condensing vapor chamber temperature was continuously monitored. The high quality video images recorded were digitally processed (using Image-J® software) to get the relevant parameters of interest, i.e., area of coverage, droplet size distribution, fall-off/slide off, and coalescence events. The primary steps in finding the area of coverage were: (a) Digital image acquisition, (b) Contrast thresholding and binning to reduce pixel noise, (c) Droplet detection with geometry attributes, (d) Measurement of total digitized pixel area covered by the droplets, and (e) Finding the area of coverage by dividing the total pixel area of all the droplets by the total area of the acquired image. Droplets below a diameter of around 0.1 mm could not be resolved with the imaging hardware used. The image processing software was first tested against benchmark images.

The experimental process was simulated by the mathematical model for both horizontal and inclined arrangement of the substrate. After validation of the simulation against experimental data, simulations were performed for the range of parameters not covered in the experiments. Here, the effect of the static contact angle, nucleation site density, thermophysical properties of the working fluid, physicochemical properties of the liquid-substrate, and the angle of inclination of the substrate are considered.

4.3.2 Experimental Validation of Computational Model

As noted, before proceeding with the simulations, benchmarking against experimental data was performed. Experimental results of condensation patterns and the corresponding predictions of numerical simulation for water vapor at a saturation temperature of 27 °C and subcooling of 5 °C are compared both in qualitative and quantitative terms underneath a horizontally oriented substarte and an inclined substrate. Nucleation sites density is taken to be 10^6 cm^{-2} in the simulation.

4.3.2.1 Horizontal Substrate

The major observable processes of dropwise condensation underneath a horizontal substrate are schematically depicted in Fig. 4.10a. These are nucleation, growth, coalescence, and fall-off of droplets; Fig. 4.10b visually and qualitatively highlights these processes, as observed experimentally (View-A in Fig. 4.9b) and captured in the computer simulation.

The statistical nature of the overall process, with multiple generations of droplets in different stages of their respective growth phase and present simultaneously on the substrate, is clearly visible. Contrary to the perfect circular footprints of the droplet bases assumed in the simulation, local phenomena such as pinning of the contact line, capillary waves, contact line inertia during droplet merging, and the dynamics of the liquid–vapor interface cause deviations that are observable in the experiments. Specifically, droplet pinning and the noncircular base of the footprint can be clearly seen in the experimental images. Thus, the mathematical model can be further refined to cover local disturbances. However, major phenomena related to dropwise condensation underneath horizontal substrates are well-simulated by the model. In Fig. 4.11 coalescence of three drops (marked a, b, and c), as observed during the experiment and revealed in the simulation, are depicted. In the simulation, the center of the new resulting drop (after coalescence, i.e., drop-d) is determined by a mass weighted average of centroid of constituent droplets before coalescence (i.e., droplets a, b, and c).

The assumption that the coalesced volume takes up the weighted center of mass of the original droplets is vindicated by this representative comparison. The merger results in the exposure of virgin areas around the drop where renucleation of the

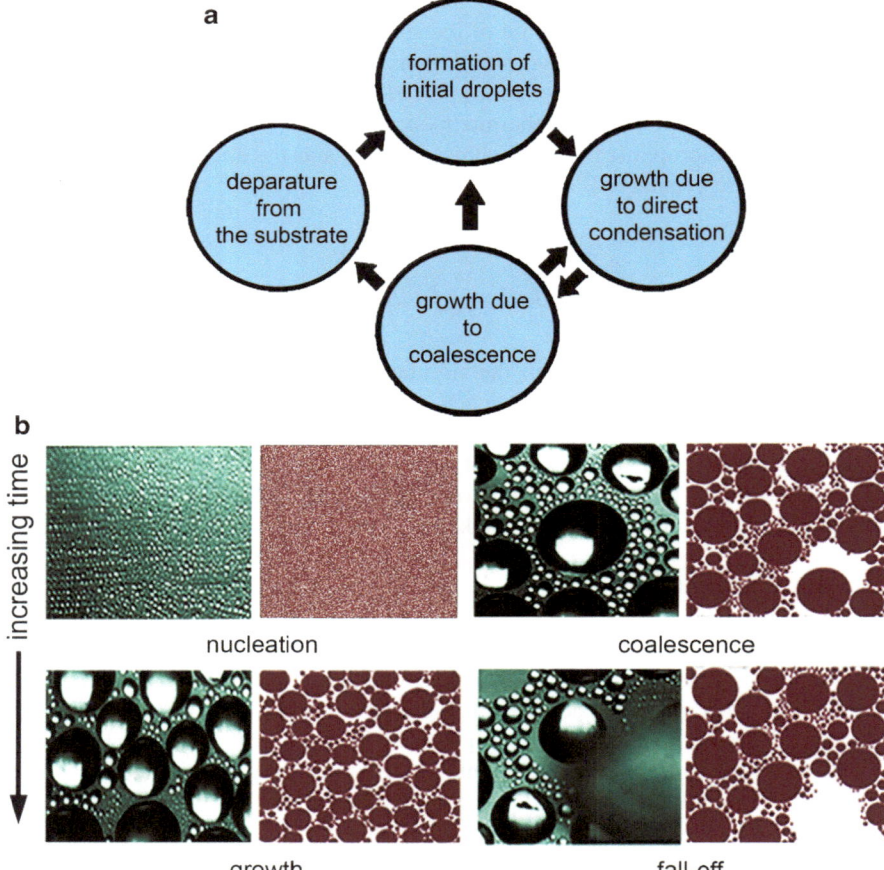

Fig. 4.10 (**a**) The cycle of major physical processes observed in the pendant mode of dropwise condensation on a horizontal substrate. (**b**) Qualitative comparison of experimental images of dropwise condensation on silanated glass substrate of area 25 mm × 25 mm (coated with octyl-decyl-tri-chloro-silane, $C_{18}H_{37}C_{13}Si$) with corresponding images generated by simulation. The hazy patch seen in the *top-left* section of the last experimental image is due to the fact that the droplet has fallen on the viewing glass through which images are recorded

new generation droplets will commence. Droplet mergers bring about near instantaneous changes in the total area coverage as well as the drop size distribution. A closer look at the edges of the droplets during experiments, especially larger droplets, also reveals that the shapes of their bases are not exactly circular, with local pinning phenomenon of the contact line occuring at certain locations (e.g., see drop d in the experimental image). As the droplets merge, experimental images show that it takes a certain finite time (of the order of 0.1–300 ms, depending on the respective sizes of coalescing droplets) for the surface and body forces to

a experiment

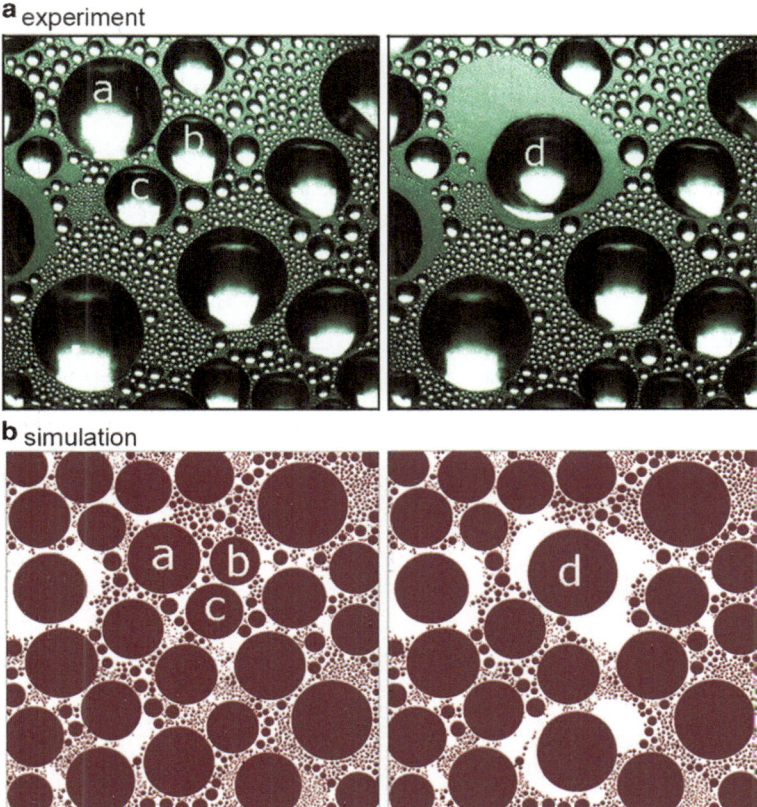

b simulation

Fig. 4.11 Sequence of two images observed during experiment (**a**) and corresponding simulation (**b**), showing coalescence of three droplets *a*, *b*, and *c*, resulting in the formation of composite drop *d*

redistribute the fluid in the coalesced drop and come to the state of minimum possible energy; the new contact line shrinks and tends to be as circular as possible in a finite relaxation time; local pinning can distort circularity.

In Fig. 4.12, the spatial drop size distributions underneath the horizontal substrate, at 15 min and 30 min respectively, after the commencement of the condensation process, is shown. No fall-off has yet taken place. The strong temporal variation of size distribution of droplets is clearly visible. As can be seen, after 15 min interval, the distribution shows moderately sized droplets with the maximum diameter of about ~2.0 mm. As the time progresses, droplets merge exposing virgin areas; an increase in number density of very small droplets (below ~0.5 mm) is clearly visible at 30 min. In addition, the number density of larger droplets (greater than ~2.0 mm) has increased substantially. The simulated histograms are denser than the experimental counterpart due to the loss of information in experimental data during image processing of droplets below about 0.1 mm. For the same

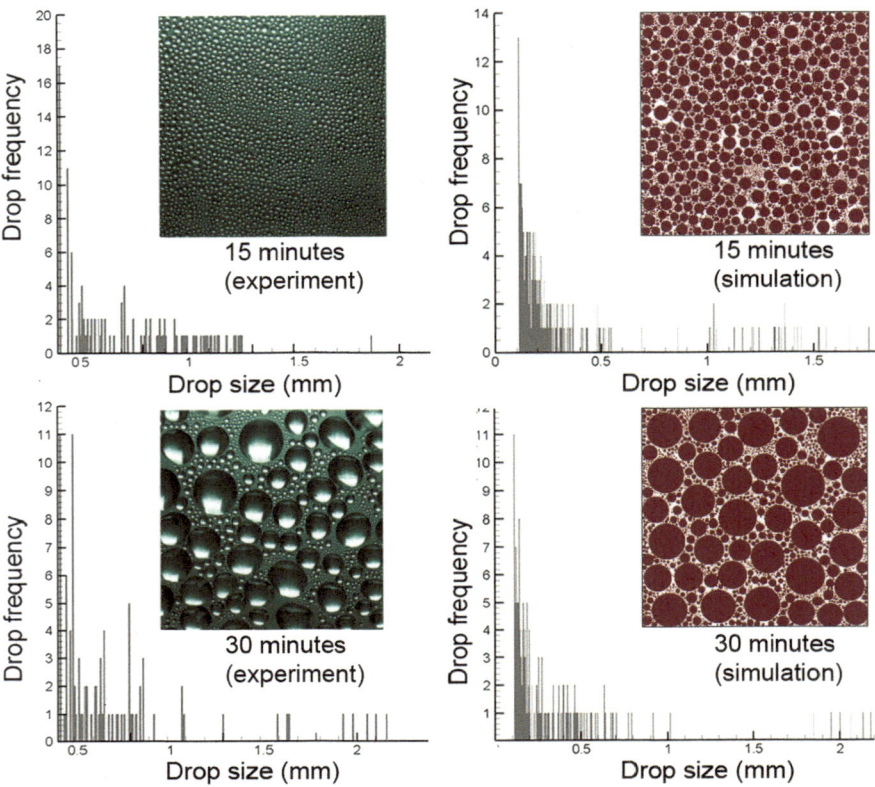

Fig. 4.12 Visual and statistical comparison of experimental and simulated spatial drop distribution patterns and the corresponding histograms of droplet frequency at the dynamic steady state

reason, the experimental and simulation histograms of the 15 min data are more dissimilar than at 30 min. Initially, as condensation commences, the number of smaller sized droplets is quite large. At later times, droplets of higher diameter are greater in number, and are captured well by the digital camera. In the latter part of the process, the growth is chiefly dominated by coalescence and the number density distribution shifts towards larger sized drops.

In Fig. 4.13a, the experimental and simulated droplet frequency plotted as a function of the drop radius, 10 min after commencement of the condensation process, are compared. The experimental fall-off time for the first drop was approximately 58–62 min while the simulation predicted a number in the range of 48–54 min. It is clear that drops whose radius is less than ~0.1 mm have not been recorded by the camera. The corresponding range of drop sizes that could be included in the simulation is 10^{-3} to 1.0 mm. Although the order of magnitude of r_{min} (at time $t = 0$) is ~10^{-4} mm, nearly all the original drops have since grown to the order of 10^{-3} mm at 10 min, mostly by direct condensation.

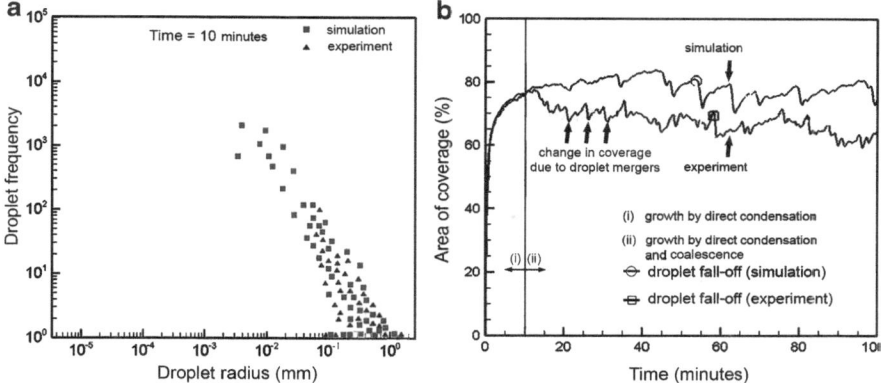

Fig. 4.13 (a) Drop size distribution from experiments and simulation at a time of 10 min after the commencement of dropwise condensation. (b) Time-wise variation of the area coverage of droplets over the substrate

Droplet coalescence has not yet started, as can be seen in Fig. 4.13b, where the temporal change in area coverage of drops is presented. Initially, there is a rapid increase in the coverage and later approaches a dynamic quasi-steady state. Two distinct zones clearly seen in the experimental and simulation data are: (a) growth due to direct condensation in the initial period and (b) growth due to coalescence. Large local fluctuations in area coverage represent time instants when drops either coalesce to form larger drops or a large drop falls-off/slides-off. The fact that smaller drops could not be accounted due to imaging limitations explains the higher values of coverage area in simulation (73.1 %) as compared to experimental data (64.5 %).

The complete sequence of experimental and simulated drop distribution, from the appearance of drops of minimum radius to the formation of drops of critical radius, underneath a horizontal substrate of 30 mm × 30 mm area, are shown in Fig. 4.14. The first image is at a time instant of 1 min and thereafter the images are at approximately 10 min intervals. The last image is presented at 65 min for the experiment and 52 min for the simulation, respectively. For this experiment, the first fall-off occurred at 59.5 min while in the corresponding simulation, the first instance of fall-off was observed at 51 min and 10 s. This discrepancy may arise due to the following factors: (a) Noncondensable gases in the experimental chamber can deteriorate the heat transfer coefficient and delay the drop growth rate. (b) Inexact nucleation site density. (c) The local effects of pinning and contact line dynamics can lead to higher frictional stresses that enhance surface forces and delay fall-off. The comparisons clearly show that the simulator satisfactorily captures the major processes of dropwise condensation, both from a qualitative and a quantitative standpoint.

a experiment

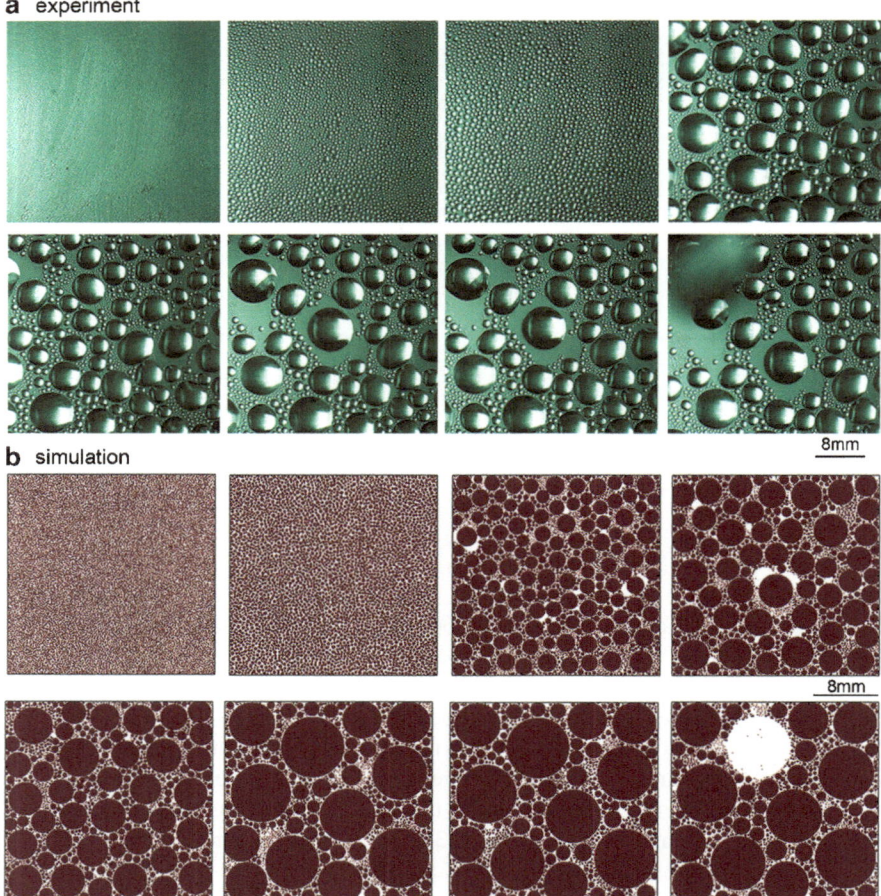

b simulation

Fig. 4.14 Comparison of experiments (**a**) and simulation (**b**) for the complete sequence of dropwise condensation process, from the appearance of drops of minimum radius to the drops of critical radius underneath a horizontal silanated glass substrate of 30 mm × 30 mm area

4.3.2.2   Inclined Substrate

Various attributes of dropwise condensation of water at saturation temperature of 27 °C ($\Delta T_{sat} = 5$ °C) underneath an inclined substrate (15° from horizontal; θ_{adv} = 111°, θ_{rcd} = 81°), recorded in experiments and observed in numerical simulation are shown in Figs. 4.15, 4.16, 4.17, and 4.18.

Major physical processes observed on an inclined substrate are similar to those of the horizontal substrate (Fig. 4.10), except that the simple fall-off mechanism is replaced by a more complex combination of slide-off and fall-off, as shown in

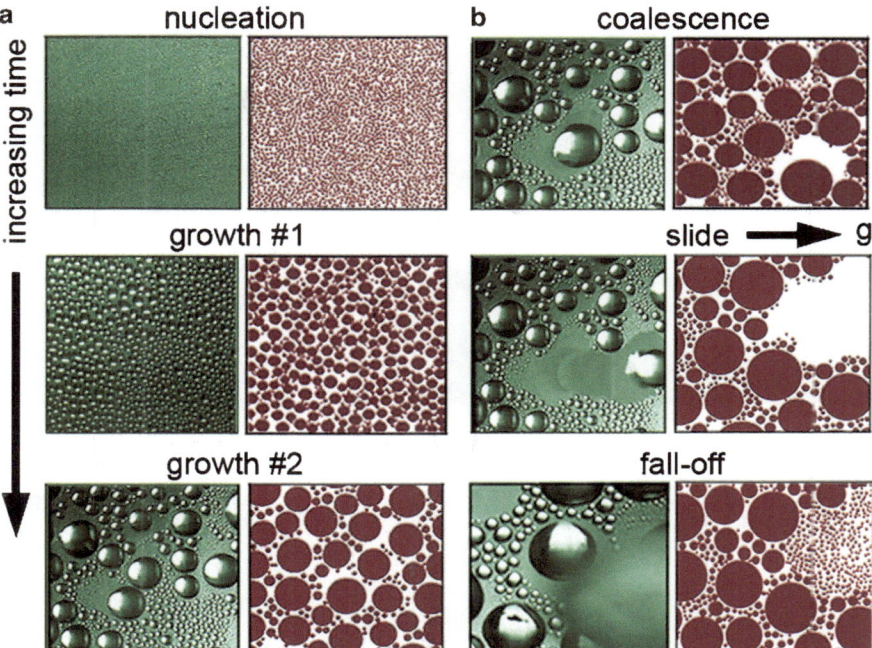

Fig. 4.15 Cycle of individual processes in dropwise condensation of vapor over an inclined cold substrate. (**a**) Cycle of individual subprocesses which constitute dropwise condensation. (**b**) Qualitative depiction of the footprints of the droplets during the cyclic growth process (for water, subcooling $\Delta T_{\text{sat}} = 5\ °C$, average contact angle = 96°, nucleation site density = $10^6\ \text{cm}^{-2}$)

Fig. 4.15. On an inclined substrate, a critically sized sliding droplet, while sweeping other droplets on its path, may either (1) reach the end of the substrate without falling off or, (2) may acquire enough mass to be pulled in the downward direction thus falling off from the substrate, before actually reaching the edge of the substrate. The scenario realized will depend on the rate of growth of the drop, coalescence, and the length of the substrate itself. The other physical processes of nucleation, direct condensation growth, coalescence, and merger dynamics are quite similar to that of horizontal substrate. The fact that the gravity vector now acts at an angle to the growing droplets leads to unsymmetrical drop deformation. The contact angle hysteresis plays a role in the static force balance.

On the inclined surface, critical sized droplets first begin to slide, rather than fall-off, as observed underneath a horizontal substrate. Criticality is achieved by direct condensation or alternatively, by coalescence with the adjoining drops. Thus, depending on the length of the substrate and timescales of direct growth and growth due to coalescence, there are various possibilities observed during the experiment underneath an inclined substrate. These include the following,

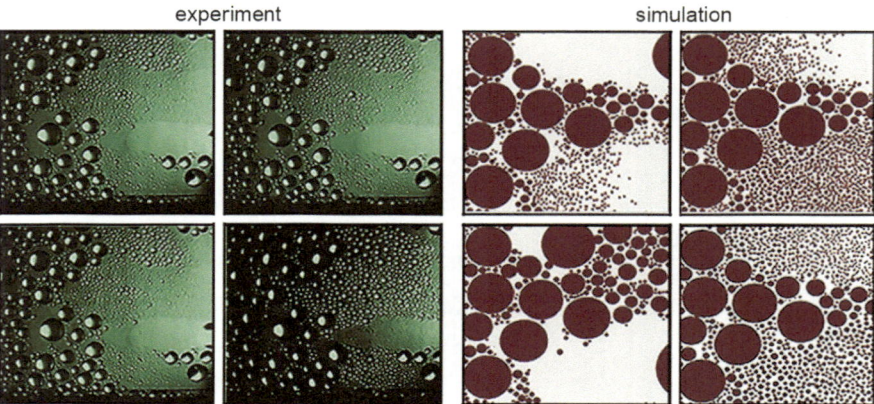

Fig. 4.16 Various stages of droplet condensation on the inclined substrate (15°) recorded during experiments and simulation. The commencement of sliding and sweeping actions of the drop as it gathers mass during transit and renucleation of the virgin exposed surface, when the sweeping action is complete, are clearly seen

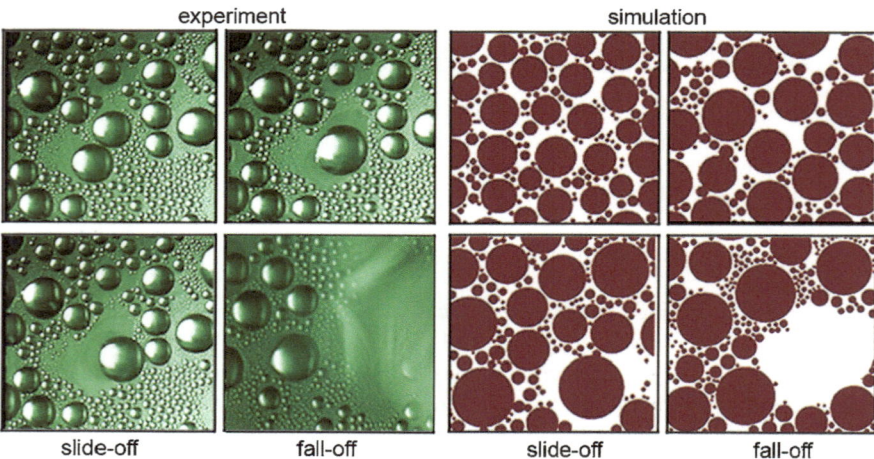

Fig. 4.17 Temporal stages of droplet condensation on the inclined substrate (5°) recorded during experiments and simulation. The sliding drop gathers mass during transit and reaches criticality of fall-off

Slide-off criticality is achieved, and during the entire slide-off on the substrate, fall-off criticality is not achieved.

Slide-off criticality is achieved, and during the slide-off underneath the substrate, fall-off critically is also achieved before the droplet traverses the complete substrate length scale. Both these possibilities have been incorporated in the simulation, as shown in Figs. 4.16 and 4.17.

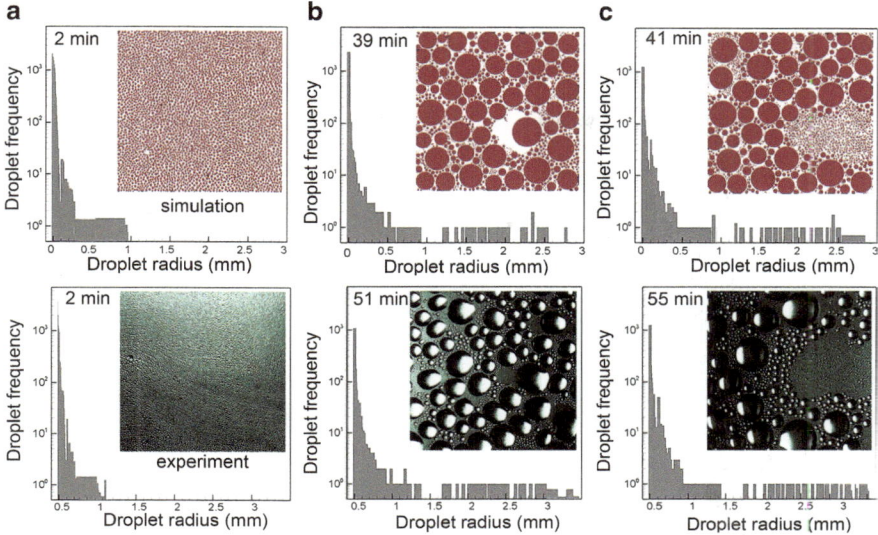

Fig. 4.18 Size distribution of drops condensing underneath an inclined (15°) silanated glass substrate of size 25 mm × 25 mm as recorded in the simulation and in experiments (**a**) at time = 2 min from the commencement of dropwise condensation, (**b**) at critical state of slide off, and (**c**) just after a complete sweeping action is completed by a sliding drop

The complex sequence of slide-off, rapidly followed by sweeping, fall-off, and renucleation, is depicted in Figs. 4.16 and 4.17. After the first instance of slide-off, it is interesting to note that the subsequent slide-offs and sweeping actions occur at a greater frequency. The mathematical model satisfactorily captures these processes.

Te experimental images and histograms of droplet frequency along with the corresponding simulation data for a hydrophobic surface of 15° inclination, are depicted in Fig. 4.18. The critical stage of slide-off is also pictorially compared; a discrepancy in the actual time of slide-off in experiments as opposed to simulation is again observed. Soon after the slide-off, virgin areas are created, fresh nucleation sites are exposed and renucleation commences, Fig. 4.18a, b.

The complete temporal sequence of events on the inclined substrate is shown in Fig. 4.19. Unlike a horizontal substrate, the drop dynamics on an inclined substrate is unique because the criticality of droplet motion and the series of events soon thereafter (sweeping and/or fall-off) happen extremely quickly leading to a sudden reduction in area coverage. Moreover, repeated removal of drops leads to the time averaged area of coverage being smaller for the inclined substrate when compared to the horizontal. At the instant of the first slide-off, the area coverage is 58.8 % in simulation and 49.5 % from experiments. The discrepancy is again attributed to the loss of data pertaining to small sized droplets during experimental observations. It is clear from Fig. 4.19 that drop slide-off underneath the inclined substrate occurs earlier than the corresponding time instant of fall-off underneath a horizontal substrate.

experiment   simulation

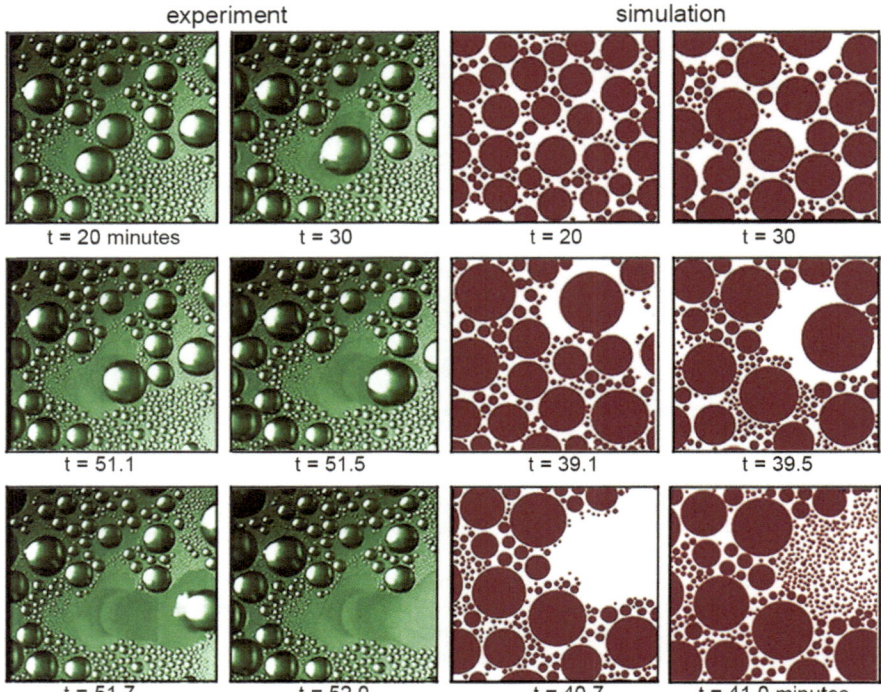

t = 20 minutes	t = 30	t = 20	t = 30
t = 51.1	t = 51.5	t = 39.1	t = 39.5
t = 51.7	t = 52.0	t = 40.7	t = 41.0 minutes

Fig. 4.19 Temporal stages of droplet condensation on an inclined substrate (15°) recorded during experiments and simulation. The commencement of sliding and sweeping actions of the drop as it gathers mass during transit and renucleation of the virgin exposed surface, when the sweeping action is complete, are clearly seen

The variation of the average substrate heat flux for dropwise condensation of water with respect to the degree of subcooling ($T_{sat} - T_w$) at condensation temperatures 30 °C and 50 °C respectively on a horizontal chemically textured substrate is shown in Fig. 4.20a. The comparison of the present simulation with the theory put forward by Le Fevre and Rose (1966) for dropwise condensation of water on a monolayer promoter layer, as reported by Rose (2002) is also shown. The model results are further validated against experiments for condensation of mercury vapor on a vertical surface (Fig. 4.20b). The experimental data for the surface-averaged wall heat flux as a function of the vapor to surface temperature difference is adopted from the work of Necmi and Rose (1977). Good overall agreement, for all levels of subcooling, is seen from Fig. 4.20. Hence, one may conclude that the model is robust and applicable to liquids, having a wide range of Prandtl numbers, condensing on substrates with various orientations and conditions.

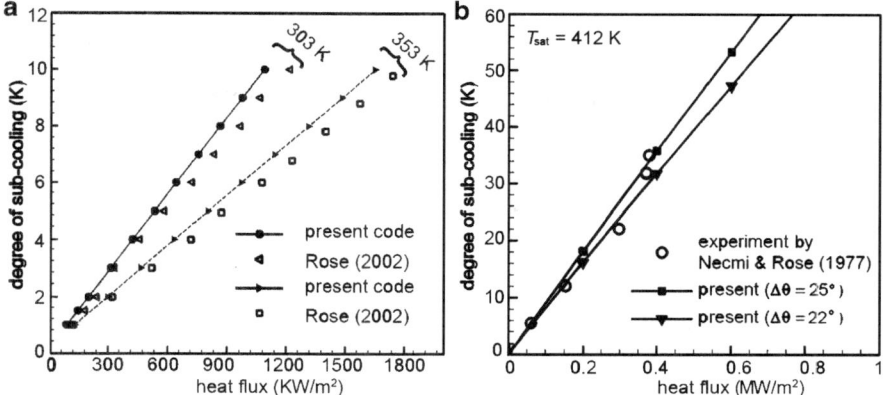

Fig. 4.20 Comparison of average substrate heat flux obtained by simulation with experimental data reported in the literature. (**a**) For water vapor condensation: Variation of heat flux, underneath a horizontal substrate, at $T_{sat} = 30$ °C and $T_{sat} = 50$ °C respectively, with the degree of subcooling. (**b**) For mercury vapor condensation: Average substrate heat flux plotted as a function of degree of subcooling (ΔT) for dropwise condensation of mercury over a vertical plate at saturation temperature of 139 °C and contact angle hysteresis ($\Delta\theta$) = 22° and 25°

4.4 Liquid Crystal Thermography of Condensing Drops

Despite sustained research in the past two decades, the prediction of the correct heat transfer rate during dropwise condensation over a surface remains a challenge (Rose 2002; Bansal et al. 2009), mainly due to lack of complete knowledge of the local transport mechanisms.

Experimental determination of the heat transfer coefficient during dropwise condensation is a demanding task because of the many intricacies involved in the process. Mainly, the driving temperature difference is small, essentially resulting in a high heat transfer coefficient. Further, uncertainties associated with the micro-scale substructure of contact line shapes and motions, dynamic temperature variations below the condensing drops, effect of roughness and inhomogeneity of the substrate structure, control of true boundary conditions, microscale instrumentation, and transport dynamics of coalescence, merger, wipe-off, renucleation cycles, and the leaching rates of the promoter layer add to the difficulty in conducting repeatable experiments. Very high heat transfer rates (and therefore a very low temperature differential) coupled with the above factors also hinder generation of repeatable experimental data. Consequently, many conflicting experimental results have been published over the years (Rose 2002). Some of the results in the literature show considerable scatter.

Although the inherent time dependence of heat transfer in dropwise condensation has been acknowledged in the literature, spatiotemporal determination of temperature fluctuations is not straightforward. Conventional thermometry (e.g., with thermocouples) cannot provide spatial information of temperature distribution. Consequently, more accurate measurements are needed to show consistency of heat transfer measurement in dropwise condensation. For this reason, Bansal et al. (2009) used liquid crystal thermography (LCT) to obtain the spatiotemporal thermal behavior of the condensing substrate. The technique was used to determine the thermal behavior at the scale of a single condensate drop. The data of Bansal et al. (2009) for measurement of heat transfer is discussed below.

Details of the experimental setup to study dropwise condensation underneath a substrate are shown in Fig. 4.21. The LCT sheet was calibrated and the calibration curve between hue and temperature is shown in Fig. 4.22. Experiments were conducted in such a way that pendant drops form on the underside of the liquid crystal sheet. The spatial distribution of temperature during dropwise condensation over a polyethylene substrate was measured using liquid crystal thermography (View B in Fig. 4.21) simultaneously with actual visualization of the condensation process using videography (View-A in Fig. 4.21). View A provides the direct picture of the drop size distribution on the substrate, whereas View B is the liquid crystal thermograph. The latter provides the hue distribution on the selected portion of the substrate as contours. These contours can be transformed into spatial temperature distribution from the hue-temperature calibration curve, as shown in Fig. 4.23.

In Fig. 4.24a, the liquid crystal thermograph of a single drop of diameter 2.96 mm, condensing on the polyethylene substrate at a vapor saturation temperature 40.3 °C, is shown. In the LCT image, regions of high heat transfer rates appear as locations of relatively high temperature; for example, the blue ring in Fig. 4.24a. Lower temperature transits towards green and red. The hue distribution over the base of the drop is shown in Fig. 4.24b.

The variation of heat throughput at the mid plane passing through the single drop, as identified in Fig. 4.24a, is shown in Fig. 4.24c. Also shown in Fig. 4.24d are examples of other isolated drops recorded during the experiments. In Fig. 4.25a, b, a pair of adjacent drops, condensing at the vapor saturation temperature of 40.3 °C and the associated hue distribution on the base area (note the elliptic shape of the droplet base), The instantaneous heat transfer rates on planes passing through these individual drops, are depicted in Fig. 4.25c, d. It is clear from the experimental images that smaller drops have a lower thermal resistance per unit area than the larger drops. Therefore, the temperature distribution is nonuniform on the condensing substrate and constriction resistance may affect heat transfer data.

In the absence of noncondensable gases, one can conclude that the periphery of the droplet base line provides the path of least resistance for heat transfer. Average heat transfer rate increases with increase in subcooling and saturation pressures. The principal finding of this study is that thermal imaging of dropwise condensation patterns over a surface is adequate for obtaining the heat transfer coefficient. These base images of the wall temperature distribution can be used to estimate local and

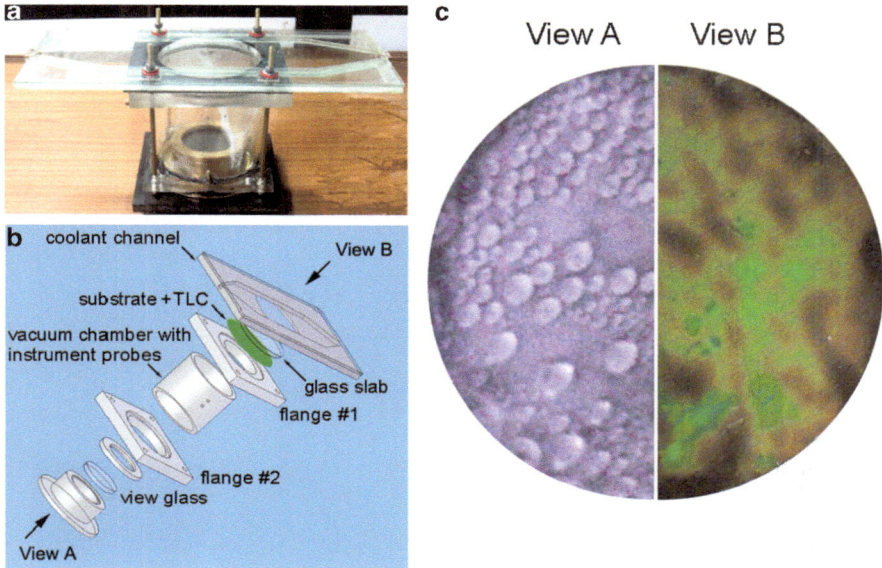

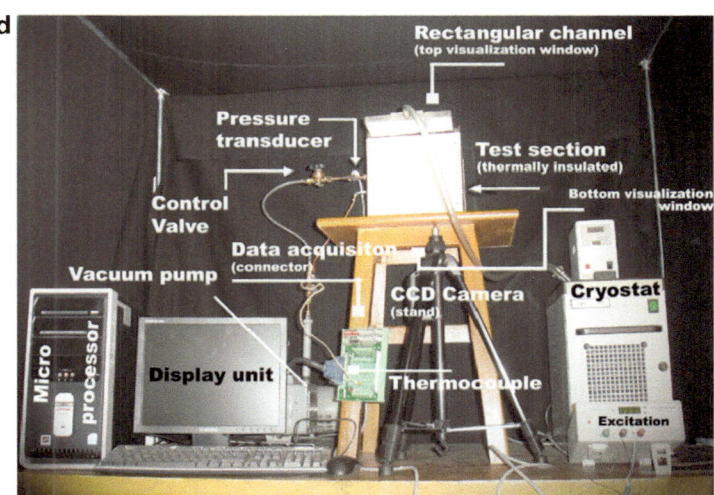

Fig. 4.21 Details of the experimental setup to study dropwise condensation on the underside of a substrate. (**a**) Photograph of the main condensing chamber; (**b**) exploded view of the chamber showing all components. (**c**) View-A (refer **b**) gives the actual photograph of condensing droplets; View-B gives RGB image of the TLC. (**d**) Photograph of the set-up (Bansal et al. 2009)

average heat transfer coefficients by including the relevant thermal resistances between the condensing vapor and the subcooled substrate.

Microscale measurement of small temperature differences continues to be a challenge for accurate estimation of heat transfer during dropwise condensation.

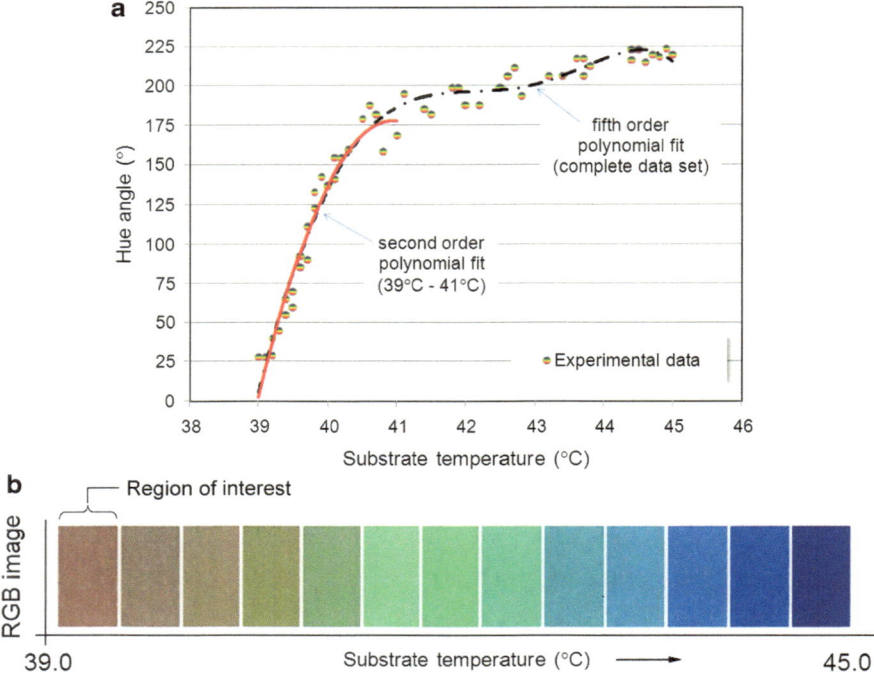

Fig. 4.22 (**a**) Calibration curve of the liquid crystal sheet relating hue to substrate temperature. (**b**) The picture shows the RGB images obtained during the calibration step. These are further processed to get the HSI images (Bansal et al. 2009)

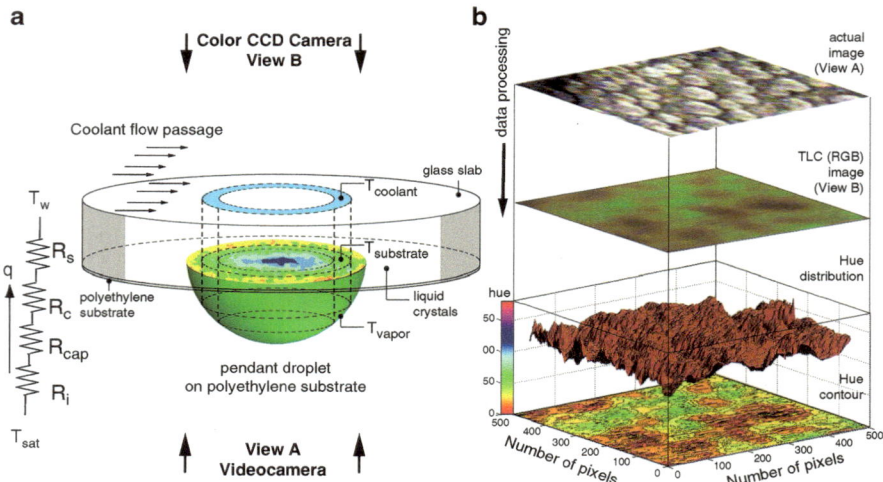

Fig. 4.23 (**a**) Overall approach for the estimation of local heat transfer coefficient for dropwise condensation occurring on the underside of a substrate. View-A provides the direct picture of the drop, whereas View-B provides the liquid crystal thermograph. (**b**) Series of operations for data reduction. The image of condensing droplets is obtained from Camera View-A, and the corresponding TLC RGB image is simultaneously obtained from Camera View-B. The latter image provides spatial hue distribution on the selected portion of the substrate, contours of which can be transformed into spatial temperature distribution from calibration curve (Bansal et al. 2009)

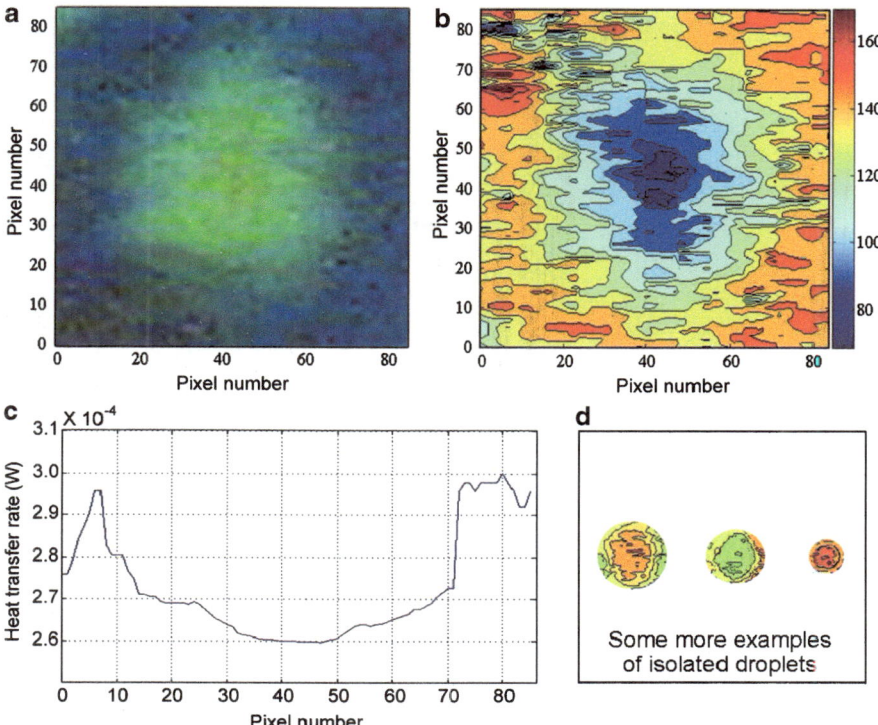

Fig. 4.24 (**a, b**) Figures show the TLC-RGB image of an isolated pendant droplet ($D = 2.96$ mm) during dropwise condensation process, at vapor saturation temperature of 40.3 °C, and its corresponding hue contour plot. Images are recorded after steady state has been attained. (**c**) The heat transfer rate at a plane passing through the middle of the droplet is shown as a function of position. (**d**) Examples of hue profiles of three other isolated droplets of smaller sizes (Bansal et al. 2009)

Although liquid crystal thermography stretches the limit of spatial resolution to microns, temperature differentials are quite small for correct and repeatable experimental determination. Numerical techniques become essential at smaller length scales.

4.5   Closure

Experiments on creating textured surfaces for dropwise condensation and the measurement of heat transfer coefficient are reviewed. Details of the experimental set-up and preparation of a chemically textured nonwetting surface for observation of dropwise condensation of water vapor underneath a horizontal and an inclined substrate are

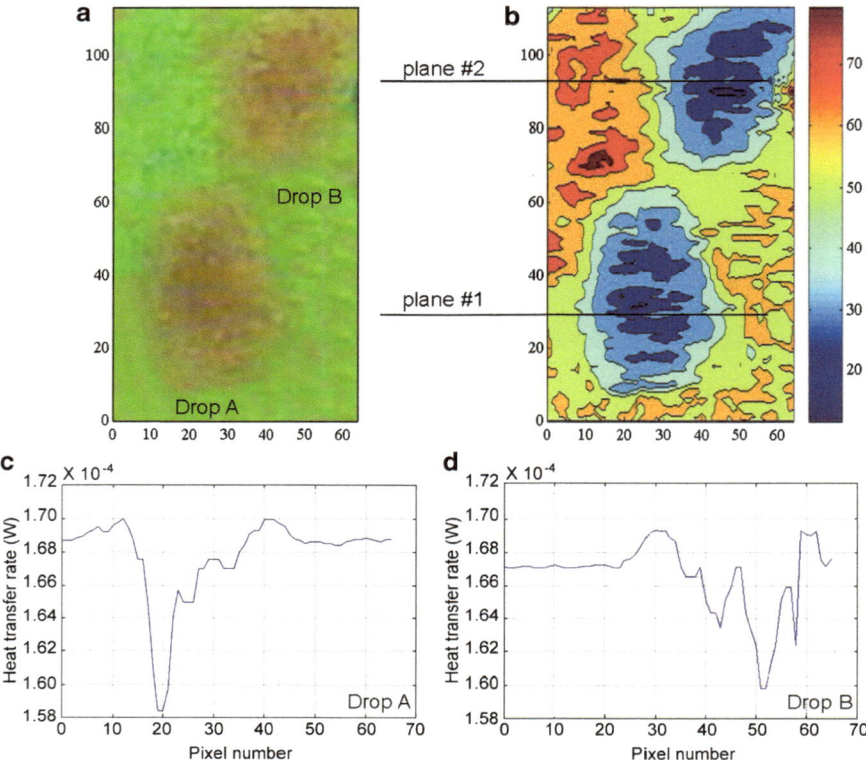

Fig. 4.25 (**a, b**) Figures show the TLC-RGB image of two adjacent pendant droplets during dropwise condensation and the corresponding hue contour plot at vapor saturation temperature of 40.3 °C. Images have been acquired at steady state. (**c**), (**d**) Heat transfer rate through *plane #1* (Drop A) and *plane #2* (Drop B) is presented as a function of position (Bansal et al. 2009)

reported. Chemical texturing of glass is achieved by silanation using octyl-decyl-tri-chloro-silane ($C_{18}H_{37}C_{13}Si$) in a chemical vapor deposition process. Experimental results of condensation patterns and the corresponding predictions of numerical simulation for water vapor are compared. The prediction of the model is in fair agreement with the experimental data of condensation of water vapor. Average heat flux as a function of degree of subcooling for water and mercury are compared. Although, there is some discrepancy in the data obtained, major phenomena related to dropwise condensation underneath horizontal substrates are well-simulated by the mathematical model.

Chapter 5
Concluding Remarks and Perspectives

Keywords Review of model developed • Comparison with experiments • Imaging coalescence • Conclusions • Scope for future work

5.1 Overview

Dropwise condensation is vapor-to-liquid phase-change in the form of discrete drops on or underneath horizontal and inclined substrates. The process is hierarchical in the sense that it occurs over a wide range of length and timescales. A mathematical model of dropwise condensation underneath textured surfaces, horizontal (with or without wettability gradient) and inclined, is reported. The model starts from the formation of drops at the atomic scale at randomized nucleation sites and follows its growth by direct condensation and coalescence, till the drop is large enough to fall-off or slide away. The atomic model shows that the largest stable cluster size in the atomic scale matches the minimum drop radius estimated from thermodynamic considerations. The minimum radius drop is insensitive to surface texturing and does not provide controllability at large length and timescales. In the model, nucleation sites are randomly distributed over the substrate. Growth rate at each nucleation site is derived on the basis that vapor condenses on the free surface of the drop and releases latent heat that is transferred through the liquid drop to the cold substrate. A simple model of coalescence has been adopted in this work. The stability criterion is developed as a force balance equation at the level of a drop. Transport parameters of a sliding drop are determined using a CFD model and presented in the form of correlations. Fluids considered are water and liquid metals such as mercury and sodium, representing a wide range of Prandtl numbers. Performing simulation of the complete cycle of dropwise condensation, the spatio-temporal distribution of drops is obtained, from which local and area-averaged heat transfer rates as a function of time are predicted.

An experimental study of water vapor condensation underneath a chemically textured substrate is carried out for validation of the complete dropwise

S. Khandekar and K. Muralidhar, *Dropwise Condensation on Inclined*
Textured Surfaces, SpringerBriefs in Applied Sciences and Technology 11,
DOI 10.1007/978-1-4614-8447-9_5, © Springer Science+Business Media New York 2014

condensation model. Substrate preparation involves coating the glass surface using chemical vapor deposition of silane molecules. The spatiotemporal drop distribution recorded during the experiment and observed in simulation underneath an inclined chemically textured substrate show fair to good agreement. Heat transfer rates are also validated against experiment data of water vapor and mercury available in the literature. Specific conclusions arrived at in the present study are listed below.

5.1.1 Drop Instability

The critical drop radius at which commencement of sliding takes place is a function of the thermophysical properties of the fluid, inclination of the substrate, and contact angle hysteresis. Fluids with higher surface tension show larger size at instability. Reduction in contact angle hysteresis reduces the critical size for a given angle of inclination.

5.1.2 Modeling Fluid Motion Inside a Moving Drop

During motion, a circulation pattern is setup within the drop. The center of the circulation pattern moves towards the solid surface at higher Reynolds numbers. Pressure and wall shear stress are nearly uniform at the base of the drop, except at the periphery where large gradients prevail. Heat transfer in drops of high Prandtl numbers is characterized by the appearance of thermal boundary layers. Temperature distribution across the drop shows large gradients near the walls while temperature inversion is seen in the core. At lower Prandtl numbers, diffusive transport governs heat transfer rates and a near-linear variation of temperature is obtained.

5.1.3 Drop Coalescence

Direct visualization shows that coalescence commences when two drops approach each other at the three-phase contact line. A tiny liquid bridge is immediately formed at the commencement of coalescence, which grows with time. Immediately afterwards, the process is limited by viscous and inertia forces, as the drop eventually attains equilibrium. The relaxation time of the composite drop is of the order of ~10–160 ms, depending on the sizes involved. This is small in comparison to the overall time period of the condensation cycle which could be several minutes long. In experiments, local pinning of the three-phase contact line is observed during coalescence while the droplet relaxes to equilibrium, in conjunction with very high dissipation at the three-phase contact line. The center of mass of the coalesced drop departs slightly from the weighted average center of the individual drops, the discrepancy being attributed to factors such as pinning. The footprint of the coalesced droplet is elliptical which eventually relaxes to a circular shape. On inclined surfaces, coalescence of drops can lead to a slide-off or fall-off.

5.1.4   Macroscopic Modeling

The overall condensation model includes the effect of contact angle, hysteresis, inclination of the substrate, thermophysical properties of fluids, nucleation site density, degree of subcooling, saturation temperature, promoter layer thickness, and wettability gradient. Simulation for various fluids and substrate inclinations shows the following trends.

1. Dropwise condensation is necessarily a quasi-cyclic process from which the average drop size, distribution, and cycle time as well as overall heat transfer coefficient and wall shear can be computed.
2. Two distinct phases of droplet growth are observed: growth due to direct condensation and growth primarily due to coalescence.
3. Increase of static contact angle (decrease in wettability of the substrate) reduces the droplet area coverage. Reduction of coverage is also observed by increasing the substrate inclination.
4. Decrease in wettability results in earlier fall-off (horizontal substrate) and earlier slide-off (inclined substrate).
5. The critical radius of droplet at which commencement of sliding takes place is a function of the thermophysical properties of the fluid, inclination of the substrate and contact angle hysteresis. Fluids with higher surface tension show larger critical radius. Reduction in contact angle hysteresis reduces the critical radius of the droplet at fall-off for a given angle of inclination.
6. Inclining the substrate results in larger number of small drops and hence in higher heat transfer coefficient.
7. Heat transfer coefficient increases with an increase in the degree of subcooling, saturation temperature, and is a strong function of the Prandtl number.
8. Providing wettability gradient serves the purpose of passively destabilizing drops in a manner similar to inclined surfaces in gravity fields. It results in a larger number of small drops and hence will lead to a higher average heat transfer coefficient.
9. The condensation cycle times of liquid metal vapor are smaller compared to water. The minimum drop radius is smaller, the maximum drop size is smaller, and area coverage is smaller, all the factors resulting in a higher heat transfer coefficient.
10. Nucleation sites density is an uncertain parameter that can be determined only indirectly but has an effect on the heat transfer coefficient. A high nucleation density leads to a high overall heat transfer coefficient.

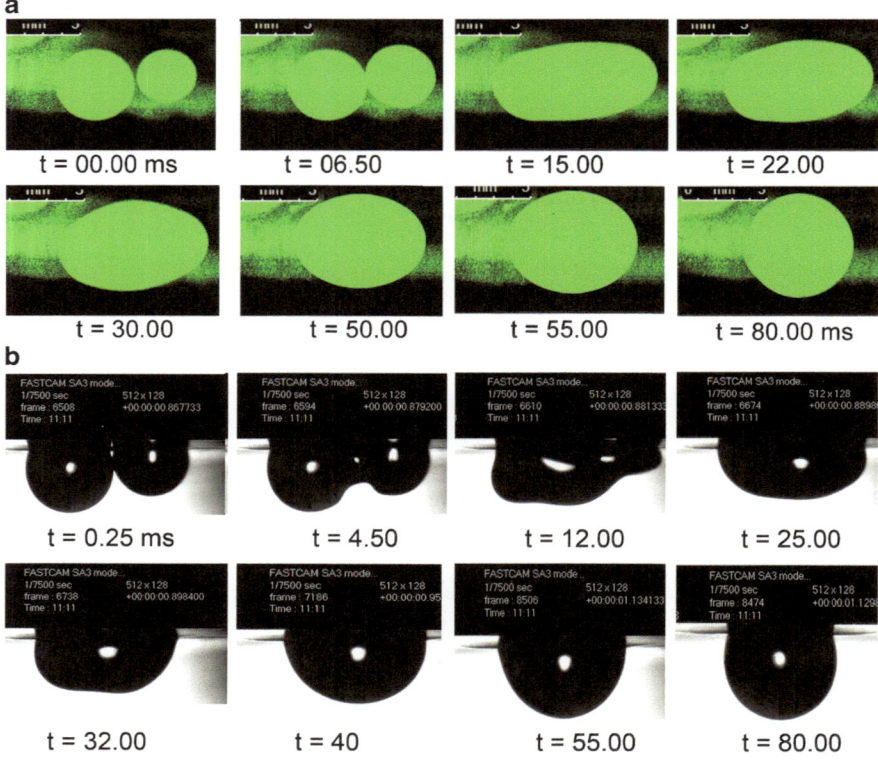

Fig. 5.1 Coalescence of two pendant droplets of water of diameters 0.85 and 1.5 mm underneath a chemically textured copper substrate (**a**) footprint of the merging drops by confocal microscope, and (**b**) snapshots recorded as a side view by high speed camera

5.1.5   Coalescence Dynamics

The coalescence model adopted in the present study approximates the nuances of the overall merger process by requiring that it occur instantaneously, moving from one equilibrium shape to the next. Preliminary experiments highlight the subtleties of the process and reveal complex flow patterns including oscillations of the free surface and large instantaneous wall shear stresses that can affect the life of the promoter layer. To address this issue, coalescence of two pendant drops of unequal volumes was carried out with a laser confocal microscope and a high-speed camera operated at 200 frames per second. Here, water is the working medium with the drops sitting under a chemically textured copper substrate. The drops were introduced one at a time from the rear side of the substrate. The length and timescales of coalescence were observed.

The image sequence recorded is summarized in Fig. 5.1. The first row of images show the evolving footprint of the merging drops as recorded by a confocal microscope. The second set of images on a grayscale is a front view of these drops as seen by a high-speed camera.

Drop coalescence commences when two droplets approach each other at the three-phase contact line at negligible velocity. A tiny liquid bridge is immediately formed, induced by the van der Waals forces. The coalescence process gets initiated by the extra surface energy released in the process. The difference in internal pressure between the two drops drives fluid motion. Immediately afterwards, the coalescence process is limited by viscous and inertia forces. In water, free surface oscillation can last ~20–40 ms, depending on the size of droplets, substrate orientation, and thermophysical properties. Long-term relaxation can occur over 40–100 ms. Rapid transients in the early stage of coalescence will induce large shear stresses over the substrate. The accompanying enhancement in heat fluxes may not compensate for the loss of life of the promoter layer.

When two or more drops coalesce underneath a horizontal substrate, the center of mass of the resultant drop at equilibrium (when motion has stopped) is located close to the weighted mass center of the individual drops. Differences may arise from pinning over portions of the three-phase contact line, particularly on inclined surfaces. Figure 5.1 shows that the footprint observed by the confocal microscope is initially an ellipsoid that relaxes in time to form a circle. Similarly, the drop shape evolves to become a part of a sphere, a trend often observed for lyophobic surfaces.

5.2   Future Work

Opting for dropwise condensation in advanced engineering systems involves not only an understanding of the fundamental thermo-fluidic transport phenomena but also the microscale issues associated with the substrate material. Hence, future research should address the following concerns:

1. Overall, hydrophobicity is desirable for dropwise condensation. Chemical texturing is limited by factors such as leaching. Creating physical textures by patterning is a long-standing problem in surface engineering. The connection of a surface pattern with contact angle and hysteresis is unresolved.
2. Experimental determination of wall shear stress and heat transfer coefficient is a challenge. The statistical nature of droplet distribution in the ensemble further contributes to the intricacy of analysis and interpretation.
3. The mathematical model presented captures the major constituents of dropwise condensation process quite satisfactorily. There are local discrepancies, however. Looking at the experiments closely, one feels that the highly localized three-phase contact line motion and the dynamics of coalescence need further investigation.
4. Information gained on the behavior of small liquid drops will be useful in several novel applications. These include microfluidic switches, bio-MEMS devices, lab-on-chip, and electro-wetting.

References

Aarts DGAL, Lekkerkerker HNW, Guo H, Wegdam GH, Bonn D (2005) Hydrodynamics of droplet coalescence. Phys Rev Lett 95:1645031–1645034

Abu-Orabi M (1998) Modeling of heat transfer in dropwise condensation. Int J Heat Mass Transf 41:81–87

Amar JG, Popescu MN, Family F (2001) Self-consistent rate-equation approach to irreversible submonolayer growth in one direction. Surf Sci 491:239–254

Andrieu C, Beysens DA, Nikolayev VS, Pomeau Y (2002) Coalescence of sessile drops. J Fluid Mech 453:427–438

Annapragada SR, Murthy JY, Garimella SV (2012) Droplet retention on an incline. Int J Heat Mass Transf 55:1457–1465

Bakulin NV, Ivanovskii MN, Sorokin VP, Subbotin VI, Chulkov BA (1967) Phase and diffusion resistance in the condensation of an alkali metal. J Atom Energy 22(5):413–415

Bansal GD, Khandekar S, Muralidhar K (2009) Measurement of heat transfer during dropwise condensation of water on polyethylene. Nanoscale Microscale Thermophys Eng 13(3):184–201

Baojin Q, Li Z, Hong X, Yan S (2011) Experimental study on condensation heat transfer of steam on vertical titanium plates with different surface energies. Exp Thermal Fluid Sci 35:211–218

Bartelt MC, Tringides MC, Evans JW (1993) Island size scaling in surface deposition processes. Phys Rev B 47:13891–13894

Bentley PD, Hands BA (1978) The condensation of atmospheric gases on cold surfaces. Proc R Soc Lond A Math Phys Sci 359:319–343

Berthier J (2008) Microdrops and digital microfluidics. William Andrew Inc., Norwich, pp 75–179

Bhushan B, Jung YC (2011) Natural and biomimetic artificial surfaces for super-hydrophobicity self-cleaning, low adhesion, and drag reduction. Progr Mater Sci 56:1–108

Bhutani G, Muralidhar K, Khandekar S (2013) Determination of apparent contact angle and shape of a static pendant drop on a physically textured inclined surface. Int Phenom Heat Transf 1:29–49

Blackman LCF, Dewar MJS, Hampson H (1957) An investigation of compounds promoting the dropwise condensation of steam. Appl Chem 7:160–157

Bloch PE, Smargiassi E, Car R, Laks DB, Andreoni W, Pantelides ST (1993) First principle calculations self-diffusion constants in silicon. Phys Rev Lett 70:2435–2438

Bonner R (2009) Condensation on surfaces with graded hydrophobicity. ASME Summer Heat Transfer Conference, San Francisco, USA

Bonner-III RW (2010) Dropwise condensation life testing of self assembled monolayers. Proceedings of the International Heat Transfer Conference (IHTC14), Washington, DC, USA

Boreyko JB, Chen CH (2009) Self-propelled dropwise condensate on superhydrophobic surfaces. Phys Rev Lett 103:184501–184504

S. Khandekar and K. Muralidhar, *Dropwise Condensation on Inclined*   133
Textured Surfaces, SpringerBriefs in Applied Sciences and Technology 11,
DOI 10.1007/978-1-4614-8447-9, © Springer Science+Business Media New York 2014

Briscoe BJ, Galvin KP (1991a) Growth with coalescences during condensation. Phys Rev A 43:1906–1918

Briscoe BJ, Galvin KP (1991b) The sliding of sessile and pendent droplets the critical condition. J Colloid Interface Sci 52:219–229

Brown RA, Orr FM, Scriven LV (1980) Static drop on an inclined plate: analysis by the finite element method. J Colloid Interface Sci 73(1):76–87

Brune H (1998) Microscopic view of epitaxial metal growth: nucleation and aggregation. Surf Sci Rep 31(3):121–229

Brune H, Bales GS, Jacobsen J, Boragno C, Kern K (1999) Measuring surface diffusion from nucleation island densities. Phys Rev B 60:5991–6006

Burnside BM, Hadi HA (1999) Digital computer simulation of dropwise condensation from equilibrium droplet to detectable size. Int J Heat Mass Transf 42:3137–3146

Carey VP (2008) Liquid-vapor phase-change phenomena, 2nd edn. Taylor and Francis Group, LLC, New York, pp 45–472

Cha TG, Yi GW, Moon MW, Lee KR, Kim HY (2010) Nanoscale patterning of micro-textured surfaces to control superhydrophobic robustness. Langmuir 26(11):8319–8326

Chaudhury MK, Whiteside GM (1992) How to make water run uphill. Science 256:1539–1541

Chen CH, Cai Q, Tsai C, Chen CL (2007) Dropwise condensation on superhydrophobic surfaces with two-tier roughness. Appl Phys Lett 90:173108–173111

Chen LH, Chen CY, Lee YL (1999) Nucleation and growth of clusters in the process of vapor deposition. Surf Sci 429:150–160

Chen LY, Baldan MR, Ying SC (1996) Surface diffusion in the low-friction limit: processes. Phys Rev B 54:8856–8861

Chen L, Liang S, Yan R, Cheng Y, Huai X, Chen S (2009) N-octadecanethiol self-assembled monolayer coating with microscopic roughness for dropwise condensation of steam. J Thermal Sci 18(2):60–165

Citakoglu E, Rose JW (1968a) Dropwise condensation some factors influencing the validity of heat-transfer measurements. Int J Heat Mass Transf 11:523–537

Citakoglu E, Rose JW (1968b) Dropwise condensation the effect of surface inclination. Int J Heat Mass Transf 12:645–451

Cras JJ, Rowe-Tait CA, Nivens DA, Ligler FS (1999) Comparison of chemical cleaning methods of glass in preparation for silanization. Biosens Bioelectron 14:683–688

Daniel S, Chaudhury MK, Chen JC (2001) Fast drop movements resulting from the phase-change on a gradient surface. Science 291:633–636

Das AK, Kilty HP, Marto PJ (2000) Dropwise condensation of steam on horizontal corrugated tubes using an organic self-asssembled monolayer coating. J Enhanced Heat Transf 7(2):109–123

Date AW (1996) Complete pressure correction algorithm for solution of incompressible Navier-Stokes equations on a nonstaggered grid. Numer heat Transf B 29:441–458

Date AW (2003) Fluid dynamical view of pressure check boarding problem and smoothing pressure correction on meshes with collocated variables. Int J Heat Mass Transf 46:4885–4898

Date AW (2005a) Introduction to computational fluid dynamics. Cambridge University Press, New York

Date AW (2005b) Solution of transport equations on unstructured meshes with cell centered collocated variables. Part I: Discretization. Int J Heat Mass Transf 48:1117–1127

DeGennes PG (1985) Wetting: static and dynamics. Rev Modern Phys 57:827–863

Dietz C, Rykaczewski K, Fedorov AG, Joshi Y (2010) Visualization of droplet departure on a superhydrophobic surface and implications to heat transfer enhancement during dropwise condensation. Appl Phys Lett 97:033104–3

Dimitrakopoulos P, Higdon JJL (1999) On the gravitational displacement of three-dimensional fluid droplets from inclined solid surfaces. J Fluid Mech 395:181–209

Duchemin L, Eggers J, Josserand C (2003) Inviscid coalescence of drops. J Fluid Mech 487:167–178

Dussan EB (1979) On the spreading of liquid on solid surfaces: static and dynamic contact lines. Annu Rev Fluid Mech 11:371–400

Dussan EB (1985) On the ability of drops or bubbles to stick to non-horizontal surface of solids. J Fluid Mech 151:1–20

Dussan EB, Chow RT (1983) On the ability of drops or bubbles to stick to non-horizontal surfaces of solids. J Fluid Mech 137:1–29

Eggers J, Lister JR, Stone HA (1999) Coalescence of liquid drops. J Fluid Mech 401:293–310

ElSherbini AI, Jacobi AM (2004a) Liquid drops on vertical and inclined surfaces I: an experimental study of drop geometry. J Colloid Interface Sci 273:556–565

ElSherbini AI, Jacobi AM (2006) Retention forces and contact angles for critical liquid drops on non-horizontal surfaces. J Colloid Interface Sci 299:841–849

ElSherbini AI, Jacobi AM (2004b) Liquid drops on vertical and inclined surfaces II: an experimental study of drop geometry. J Colloid Interface Sci 273:566–575

Erb RA (1965) Promoting permanent dropwise condensation. Ind Eng Chem 57:49–52

Erb RA (1973) Dropwise condensation on gold. Gold Bull 6:2

Erb RA, Thelen E (1965) Dropwise condensation on hydrophobic metal and metal-sulfide surfaces. 149th National meeting and Symposium of the American Chemical Society, Detroit, Michigan, USA

Erb RA, Thelen E (1966) Dropwise condensation characteristics of permanent hydrophobic system. U. S. Department of Interior, R&D Report # 184. pp. 5–57

Eucken A (1937) Energie und stoffaustauch an grenzflaechen. Naturwissenschaften 25:209–219

Extrand CW, Kumagai Y (1995) Liquid drop on an inclined plane: the relation between contact angles drop shape and retentive forces. J Colloid Interface Sci 170:515–521

Fang C, Hidrovo C, Wang F, Eaton J, Goodson K (2008) 3-D numerical simulation of contact angle hysteresis for microscale two phase flow. Int J Multiphas Flow 34:690–705

Feng J, Qin Z, Yao S (2012) Factors affecting the spontaneous motion of condensate drops on superhydrophobic copper surfaces. Langmuir 28:6067–6075

Furmidge CG (1962) The sliding drop on solid surfaces and a theory for spray retention. J Colloid Sci 17:309–324

Gao L, McCarthy TJ (2006) Contact angle hysteresis explained. Langmuir 22:6234–6237

Genzer J, Efimenko K (2000) Creating long-lived super hydrophobic polymer surface through mechanically assembled monolayer. Science 290:2130–2133

Glicksman RL, Hunt WA (1972) Numerical simulation of dropwise condensation. Int J Heat Mass Transf 15:2251–2269

Gose E, Mucciordi AN, Baer E (1967) Model for dropwise condensation on randomly distributed sites. Int J Heat Mass Transf 10:15–22

Graham C (1969) The limiting heat transfer mechanism of dropwise condensation. PhD thesis, Massachusetts Institute of Technology, USA

Graham C, Griffith P (1973) Dropwise size distribution and heat transfer in dropwise condensation. Int J Heat Mass Transf 16:337–346

Grand NL, Daerr A, Limit L (2005) Shape and motion of drops sliding down an inclined plane. J Fluid Mech 541:253–315

Greenspan HP (1978) On the motion of a small viscous droplet that wets a surface. J Fluid Mech 84(1):125–143

Griffith P (1985) Dropwise condensation. In: Rohsenow WP, Hartnett JP, Ganic EN (eds) Handbook of heat transfer fundamentals, 2nd edn. Mcgraw-Hill, New York, pp 37–49

Griffith P, Lee MS (1967) The effect of surface thermal properties and finish on dropwise condensation. Int J Heat Mass Transf 10:697–707

Grischke M, Trojan K, Dimigen H (1994) Deposition of low energy coating with dLC-like properties. Proceedings of 11th conference on high vacuum, interfaces and thin films. Dresden, Germany, pp 433–436

Gu Y, Liao Q, Zhu X, Wang H (2005) Dropwise condensation heat transfer coefficient on the horizontal surface with gradient surface energy. J Eng Thermophys 26(5):820–822

Hao P, Lv C, Yao Z, He F (2010) Sliding behavior of water droplet on superhydrophobic surface. Lett J Explor (EPL) 90:660031–660036

Hashimoto H, Kotake S (1995) In-situ measurement of clustering process near condensate. Thermal Sci Eng 3:37–43

Hatamiya S, Tanaka H (1986) A study on the mechanism of dropwise condensation (1st Report, Measurement of Heat-Transfer Coefficient of Steam at Low Pressures). Trans JSME Ser B 52(476):1828–1833

Hsieh CT, Chen WY, Wu FL (2008) Fabrication and super-hydrophobicity of fluorinated carbon fabrics with micro/nano-scaled two-tier roughness. Carbon 46:1218–1224

Huh C, Mason SG (1977) Effect of surface roughness on wetting (theoretical). J Colloids Interface Sci 60:11–37

Ivanovskii MN, Subbotin VI, Milovanov YV (1967) Heat transfer with dropwise condensation of mercury vapor. Teploenergetika 14:81–85

Izumi M, Kumagai S, Shimada R, Yamakawa N (2004) Heat transfer enhancement of dropwise condensation on a vertical surface with round shaped grooves. Exp Therm Fluid Sci 28(2–3):243–248

Jakob M (1936) Heat transfer in evaporation and condensation. Mech Eng 58:643–660

Jessensky O, Müller F, Gösele U (2003) Self-organized formation of hexagonal pore arrays in anodic alumina. Appl Phys Lett 72(9):1173–1175

Kapur N, Gaskell PH (2007) Morphology and dynamics of droplet coalescence on a surface. Phys Rev Lett 97:0563151–0563154

Kim HY, Lee H, Kang BH (2002) Sliding of drops down an inclined solid surface. J Colloid Sci 247:372–382

Kim S, Kim KJ (2011) Dropwise condensation suitable for superhydrophobic surfaces. ASME J Heat Transf 133(8):0815021–0815028

Koch G, Zhang DC, Leiertz A (1997) Condensation of steam on the surface of hard coated copper discs. Heat Mass Transf 32:149–297

Koch G, Zhang D, Leipertz A (1998) Study of plasma enhanced CVD coated material to promote dropwise condensation. Int J Heat Mass Transf 41(13):1899–1900

Kotake S (1998) Molecular clusters. In: Tien C-L, Majumdar A, Gerner FM (eds) Microscale energy transport. Taylor & Francis, Washington, DC, pp 167–185

Krasovitski B, Marmur A (2005) Drops down the hill: theoretical study of limiting contact angles and the hysteresis range on a tilted plate. Langmuir 21(9):3881–3885

Krischer S, Grigull U (1971) Microscopic study of dropwise condensation. Wärme-und Stoffübertragung 4:48–59

Lan Z, Ma XZ, Zhang Y, Zhou XD (2009) Theoretical study of dropwise condensation heat transfer: effect of the liquid-solid surface free energy difference. J Enhanc Heat Transf 16:61–71

Lara JR, Holtzapple MT (2011) Experimental investigation of dropwise condensation on hydrophobic heat exchangers. Part II: effect of coatings and surface geometry. Desalination 280:363–369

Lau KKS, Bico J, Teo KBK, Chowilla M, Amaratunga GAJ, Milne WI, McKinley GH, Gleason KK (2003) Superhydrophobic carbon nanotube forests. Nano Lett 3:1701–1705

Lawal A, Brown RA (1982) The stability of an inclined pendent drop. J Colloid Interface Sci 89:332–345

Le Fevre EJ, Rose JW (1965) An experimental study of heat transfer by dropwise condensation. Int J Heat Mass Transf 8:1117–1133

Le Fevre EJ, Rose JW (1966) A theory of heat transfer by dropwise condensation. Proc. 3rd international heat transfer conference, Chicago, vol. 2. pp. 362–375

Le Fevre EJ, Rose JW (1964) Heat-transfer measurement during dropwise condensation of Steam. Int J Heat Mass Transf 7:272–273

Leach RN, Stevens F, Langford SC, Dickinson JT (2006) Dropwise condensation: experiments and simulations of nucleation and growth of water drops in a cooling System. Langmuir 22:8864–8872

Lee LY, Fang TH, Yang YM, Maa JR (1998) The enhancement of dropwise condensation by wettability modification of solid surface. Int Commun Heat Mass Transf 25(8):1095–1103

Lee YL, Maa JR (1991) Nucleation and growth of condensate clusters on solid surfaces. J Mater Sci 26:6068–6072

Leger L, Joany JF (1977) Liquid spreading. Rep Progr Phys 57:431–487

Leipertz A (2010) Dropwise condensation. In: Stephan P (ed) VDI heat atlas VDI-GVC, 2nd edn. Springer, Germany, pp 933–937

Leipertz A, Cho KH (2000) Dropwise condensation on ion implanted metallic surfaces. Proceedings of the third European thermal sciences conference. pp. 917–921

Leipertz A, Fröba AP (2006) Improvement of condensation heat transfer by surface modification. Proceedings of the seventh ASME, heat and mass transfer conf., IIT Guwahati, India, K7. pp. k85–k99

Leipertz A, Fröba AP (2008) Improvement of condensation heat transfer by surface modifications. Heat Transf Eng 29(4):343–356

Lenz P, Lipowdky R (1998) Morphological transitions of wetting layer on structured surfaces. Phys Rev Lett 80(9):1920–1998

Li W, Amirfazli A (2007) Microtextured superhydrophobic surfaces: a thermodynamic analysis. Adv Colloid Interface Sci 132:51–68

Liao Q, Wang H, Zhu X, Li M (2006) Liquid droplet movement on horizontal surface with gradient surface energy. Sci Chin Ser E Technol Sci 49(6):733–741

Liao Q, Zhu X, Xing SM, Wang H (2008) Visualization study on coalescence between pair of water drops on inclined surfaces. Exp Thermal Fluid Sci 32(8):1647–1654

Liu T, Mu C, Sun X, Xia S (2007) Mechanism study on formation of initial condensate droplets. AIChE J 53(4):1050–31055

Liu Y, Chen X, Xin JH (2006) Super-hydrophobic surfaces from a simple coating method: a bionic nanoengineering approach. Nanotechnology 17:3259–3263

Ma XH, Zhou XD, Lan Z, Li YM, Zhang Y (2008) Condensation heat transfer enhancement in the presence of non-condensable gas using the interfacial effect of dropwise condensation. Int J Heat Mass Transf 51:1728–1737

Ma X, Wang B (1999) Life time test of dropwise condensation on polymer-coated surfaces. Heat Transf Asian Res 28(7):551–558

Ma X, Chen J, Xu D, Lin J, Ren C, Long Z (2002) Influence of processing conditions of polymer film on dropwise condensation heat transfer. Int J Heat Mass Transf 45:3405–3411

Ma X, Rose JW, Xu D, Lin J, Wang B (2000a) Advances in dropwise condensation heat transfer. Chin Res Chem Eng J 78:78–93

Ma X, Tao B, Chen J, Xu D, Lin J (2000b) Dropwise condensation heat transfer of steam on a polytethefluoroethylene film. J Thermal Sci 10(3):247–253

Ma X, Wang S, Lan Z, Peng B, Ma HB, Cheng P (2012) Wetting mode evolution of steam dropwise condensation on superhydrophobic surface in the presence of non-condensable gas. ASME J Heat Transf 134:021501–021509

Maa JR (1978) Drop-size distribution and heat flux of dropwise condensation. Chem Eng J 16:171–176

Majumdar A, Mezic I (1999) Instability of ultra-thin water film and the mechanism of droplet formation on hydrophobic surfaces. J Heat Transf 121:964–970

Mareka R, Straub J (2001) Analysis of the evaporation coefficient and the condensation coefficient of water. Int J Heat Mass Transf 44:39–53

Marto PJ, Looney DJ, Rose JW (1986) Evaluation of organic coating for the promotion of dropwise condensation of steam. Int J Heat Mass Transf 29:1109–1117

McCarthy M, Enright R, Gerasopoulos K, Culver J, Ghodssi R, Wang EN (2010) Biomimetic superhydrophobic surfaces using viral nano templates for self-cleaning and drop-wise condensation. Proc. of the 2010 solid state sensor actuator and micro-system workshop, Hilton Head, SC, USA

McCormic JL, Baer E (1963) On the mechanism of heat transfer in dropwise condensation. J Collide Sci 18:208–216

McCormick JL, Westwater JW (1965) Nucleation sites for dropwise condensation. Chem Eng Sci 20:1021–1031

McCoy BJ (2000) Vapor nucleation and droplet growth: cluster distribution kinetics for open and closed Systems. J Colloid Interface Sci 228:64–72

Meakin P (1992) Steady state behavior in a model for droplet growth sliding and coalescence: the final stage of dropwise condensation. Phys A 183:422–438

Mikic BB (1969) On the mechanism of dropwise condensation. Int J Heat Mass Transf 12:1311–1315

Miljkovic N, Enright R, Wang EN (2012) Effect of droplet morphology on growth dynamics and heat transfer during condensation on superhydrophobic nanostructured surfaces. ACS Nano 6:1776–1785

Mills AF, Seban RA (1967) The condensation coefficient of water. Int J Heat Mass Transf 10:1815–1827

Miwa M, Nakajima A, Fujishima A, Kazuhito HK, Toshiya Watanabe T (2000) Effects of the surface roughness on sliding angles of water droplets on superhydrophobic surfaces. Langmuir 16:5754–5760

Mori K, Fujita N, Horie H, More S, Miyashita M, Matsuda M (1991) Heat transfer promotion of aluminum–brass cooling tube by surface treatment with triazinethiols. Langmuir 7:1161–1166

Moumen N, Subramanian R, McLaughlin JB (2006) Experiments on the motion of drops on a horizontal solid surface due to wettability gradient. Langmuir 22:2682–2690

Mu C, Pang J, Liu T (2008) Effect of surface topography of material on nucleation site density of dropwise condensation. Chem Eng Sci 63:874–880

Nakajima A, Hashimoto K, Watanabe T (2001) Recent studies on super-hydrophobic films. Monatschefte Chem 132:31–41

Narhe R, Beysens D, Nikolayev VS (2004) Contact line dynamics in drop coalescence and spreading. Langmuir 20:1213–1221

Narhe R, Beysens D, Nikolayev VS (2005) Dynamics of drop coalescence on a surface: the role of initial conditions and surface properties. Int J Thermophys 26:8593–8597

Necmi S, Rose JW (1977) Heat-transfer measurements during dropwise condensation of mercury. Int J Heat Mass Transf 20:877–880

Neumann AW, Abdelmessih AH, Hameed A (1978) The role of contact angles and contact angles hysteresis in dropwise condensation heat transfer. Int J Heat Mass Transf 21:947–953

Niknejad J, Rose JW (1984) Comparisons between experiment and theory for dropwise Condensation. Int J Heat Mass Transf 20:2253–2257

Ondarçuhu T (1995) Total or partial pinning of a droplet on a surface with chemical discontinuity. J Phys II France 5:227–241

Öner D, McCarthy TJ (2000) Ultra-hydrophobic surfaces: effects of topography length scales on wettability. Langmuir 16:7777–7782

Oura K, Lifshits VG, Saranin AA, Zotov AV, Katayama M (2003) Surface science, 1st edn. Springer, Berlin, pp 220–260

Paulsen JD, Burton JC, Nagel SR (2011) Viscous to inertial crossover in liquid drop coalescence. Phys Rev Lett 106:1145011–1145014

Peng XF, Liu D, Lee DJ, Yan Y, Wang BX (2000) Cluster dynamics and fictitious boiling in micro-channels. Int J Heat Mass Transf 43:4259–4265

Pierce E, Carmona FJ, Amirfazli A (2008) Understanding of sliding and contact angles results in tilted plate experiments. Colloid Surfaces-A Physicochem Eng Aspects 323:73–82

Pozrikidis C (2009) Fluid dynamics: theory, computation, and numerical simulation, 2nd edn. Springer, New York, pp 237–250

Pratap V, Moumen N, Subramanian R (2008) Thermocapillary motion of a liquid drop on a horizontal solid surface. Langmuir 24:2185–5193

Ratsch C, Venables JA (2003) Nucleation theory and the early stages of thin film growth. J Vacuum Soc Technol A21(5):s96–s109

Ratsch C, Zangwill A (1994) Saturation and scaling of epitaxial island densities. Phys Rev Lett 72:3194–3197

Ratsch C, Seitsonen AP, Scheffler M (1997) Strain dependence of surface diffusion: Ag on Ag (111) and Pt (111). Phys Rev B 55:6750–6753

Rausch MH, Fröba AP, Leipertz A (2007) Dropwise condensation on plasma-ion implanted aluminum surface. Int J Heat Mass Transf 51:1061–1070

Rausch MH, Leipertz A, Fröba AP (2010a) Dropwise condensation of steam on ion implanted titanium surfaces. Int J Heat Mass Transf 53:423–430

Rausch MH, Leipertz A, Fröba AP (2010b) On the mechanism of dropwise condensation on ion implanted metallic surface. ASME J Heat Transf 132:945031–945033

Rio E, Daerr A, Andreotti B, Limat L (2005) Boundary conditions in the vicinity of a dynamic contact line: experimental investigation of viscous drops sliding down an inclined plane. Phys Rev Lett 94:0245031–0245034

Ristenpart WD, McCalla PM, Roy RV, Stone HA (2006) Coalescence of spreading droplets on a wettable substrate. Phys Rev Lett 97:0645011–0645014

Rose JW (1972) Dropwise condensation of mercury. Int J Heat Mass Transf 15:1431–1434

Rose JW (1976) Further aspects of dropwise condensation theory. Int J Heat Mass Transf 19:1363–1370

Rose JW (1978a) The effect of surface thermal conductivity on dropwise condensation heat transfer. Int J Heat Mass Transf 21:80–81

Rose JW (1981) Condensation theory. Int J Heat Mass Transf 24:191–194

Rose JW (1998) Condensation heat transfer fundamentals. Trans AIChE 76A:143–152

Rose JW (2004) Surface tension effects and enhancements of condensation heat transfer. Chem Eng Res Des 82:419–429

Rose JW, Glicksman LR (1973) Dropwise condensation-the distribution of drop sizes. Int J Heat Mass Transf 16:411–425

Rose J, Utaka Y, Tanasawa I (1999) Dropwise condensation. In: Kandlikar SG (ed.) Handbook of phase change: boling and condensation. Taylor and Francis, USA. pp. 581–594

Rose JW (1978b) Effect of tube conductivity material on heat transfer during dropwise condensation of steam. Int J Heat Mass Transf 21:835–840

Rose JW (2002) Dropwise condensation: theory and experiments: a review. Proc Instit Mech Eng UK 216:115–118

Rykaczewski K (2012) Microdroplet growth mechanism during water condensation on superhydrophobic surfaces. Langmuir 28:7720–7729

Sakai M, Hashimoto A (2007) Image analysis system for evaluating sliding behavior of a liquid droplet on a hydrophobic surface. Rev Sci Instrum 78:045103–045109

Sakai M, Hashimoto A, Yoshida N, Suzuki S, Kameshima Y, Nakajima A (2006) Direct observation of internal fluidity in a water droplet during sliding on hydrophobic surfaces. Langmuir 22:4906–4909

Schmidt E, Schurig W, Sellschopp W (1930) Versuche über die condensation von wasserdampf in film- und tropfenform. Forsch Ingenieurwes 1(2):53–63

Sellier M, Trelluyer E (2009) Modeling the coalescence of sessile droplets. Biomicrofluidics 3:0224121–02241213

Sellier M, Nock V, Verdier C (2011) Self-propelling coalescing droplets. Int J Multiphas Flow 37:462–468

Shi F, Shim Y, Amar JG (2005) Island-size distribution and capture numbers in three-dimensional nucleation: comparison with mean-field behavior. Phys Rev B 71:245411–245416

Shibuichi S, Onda T, Satoh N, Tsujii K (1996) Super-water-repellent fractal surfaces. J Phys Chem 100:19512–19617

Sikarwar BS, Battoo NK, Khandekar S, Muralidhar K (2011) Dropwise condensation underneath chemically textured surfaces: simulation and experiments. ASME J Heat Transf 133(2):0215011–02150115

Sikarwar BS, Khandekar S, Muralidhar K (2013a) Mathematical modeling of dropwise condensation on textured surfaces. Sadhana (Springer), Special issue on Multi-phase flows with phase-change accepted for publication

Sikarwar BS, Khandekar S, Muralidhar K (2013) Simulation of flow and heat transfer in a liquid drop sliding underneath a hydrophobic surface. Int J Heat Mass Transf 57(2):786–811

Sikarwar BS, Khandekar S, Agrawal S, Kumar S, Muralidhar K (2012) Dropwise condensation studies on multiple scales. Heat Transf Eng Special Issue Adv Heat Transf 33(4–5):301–341

Sommers AD, Jacobi AM (2006) Creating micro-scale surface topology to achieve anisotropic wettability on an aluminum surface. J Micromechan Micro Eng 16:1571–1578

Sommers AD, Jacobi AM (2008) Wetting phenomena on micro-grooved aluminum surfaces and modeling of the critical droplet size. J Colloid Interface Sci 328:402–411

Song T, Lan Z, Ma X, Bai T (2009) Molecular clustering physical model of steam condensation and the experimental study on the initial droplet size distribution. Int J Thermal Sci 48:2228–2236

Song Y, Xu D, Lin J, Tsian S (1991) A study on the mechanism of dropwise condensation. Int J Heat Mass Transf 34(11):2827–2831

Stephan K (1992) Heat transfer in condensation and boiling. Springer, Berlin. pp. 28–77

Stroscio JS, Pierce DT (1994) Scaling of diffusion mediated island growth in iron-on-iron homoepitaxy. Phys Rev B 49:8522–8525

Stylianou SA, Rose JW (1983) Drop-to-filmwise condensation transition: heat transfer measurements for ethandiol. Int J Heat Mass Transf 26(5):747–760

Stylianou SA, Rose JW (1980) Dropwise condensation on surface having different thermal conductivities. ASME J Heat Transf 102:477–482

Subramanian R, Moumen N, McLaughlin JB (2005) The motion of a drop on a solid surface due to a wettability gradient. Langmuir 21:11844–11849

Sukhatme SP, Rohsenow WM (1966) Heat transfer during film condensation of a liquid metal vapor. ASME J Heat Transf 88:19–28

Suzuki S, Nakajima A, Sakai M, Song J, Yoshida N, Kameshima Y, Okada K (2006) Sliding acceleration of water droplets on a surface coated with fluoroalkysline and octadecyltri-methoxysilane. Surf Sci 600:2214–2219

Takeyama T, Shimizu S (1974) On the transition of dropwise-film condensation. Proc 5th Int Heat Transf Conf 3:274–278

Tanaka H (1975a) A theoretical study of dropwise condensation. J Heat Transf 97(1):97–103

Tanaka H (1975b) Measurement of drop-size distribution during transient dropwise condensation. J Heat Transf 97:341–346

Tanasawa I, Utaka Y (1983) Measurement of condensation curves for dropwise condensation of steam at atmospheric pressure. J Heat Transf 1(05):633–638

Tanasawa I (1991) Advances in condensation heat transfer. In: Hartnett JP, Irvine TF, Cho IY (eds.) Advances in heat transfer, vol. 21. pp. 56–136

Tanasawa I, Ochiai J, Utaka Y, Enya S (1976) Experimental study on dropwise condensation (Effect of departing drop size on heat-transfer coefficient). Trans JSME 42(361):2846–2853

Tanasawa I, Ochiiai J, Utaka Y, En-Ya S (1974) Dropwise condensation. 11th Japanese heat transfer symposium, vol. 229

Tanner DW, Pope D, Potter CJ, West D (1968) Heat transfer in dropwise condensation at low steam pressure in the absence of non-condensable gas. Int J Heat Mass Transf 11:181–190

Thoroddsen ST, Qian B, Etoh TG, Takehara K (2007) The initial coalescence of miscible drops. Phys Fluids 19:0721101–07211020

Thoroddsen ST, Takehara K, Etoh TG (2005) The coalescence speed of pendent and a sessile drop. J Fluid Mech 527:85–114

Tian Y, Wang XD, Peng XF (2004) Analysis of surface inside metastable bulk phase during gas–liquid phase transition. J Eng Thermophys (Chin) 25:100–102

Trimmer WSN (1989) Microrobots and micromechanical system. Sensors Actuators 19:267–287

Tsuruta T (1993) Constriction resistance in dropwise condensation. Proc. of the ASME engineering foundation conference on condensation and condenser design. pp. 109–170

Ucar IO, Erbil HY (2012) Dropwise condensation rate of water breath figures on polymer surfaces having similar surface free energies. Appl Surf Sci 259:515–523

Umur A, Griffith P (1965) Mechanism of dropwise condensation. ASME J Heat Transf 87:275–282

Utaka Y, Kubo R, Ishii K (1994) Heat transfer characteristics of condensation of capor on a lyophobic Surface. Proc 10th Int Heat TransfConf 3:401–406

Utaka Y, Saito A, Ishikawa H, Yanagida H (1987) Transition from dropwise condensation to film condensation of propylene glycol ethylene glycol, and glycerol vapors. Proc 2nd ASME-1SME Thermal Eng Conf 4:377–384

Vemuri S, Kim KJ (2006) An experimental and theoretical study on the concept of dropwise condensation. Int J Heat Mass Transf 49:649–857

Vemuri S, Kim KJ, Wood BD, Govindaraju S, Bell TW (2006) Long term testing for dropwise condensation using self-assembled monolayer coating of n-octadecyl mercaptan. Appl Thermal Eng 26:421–429

Venables JA (2000) Introduction to surface and thin film processes, 1st edn. Cambridge University Press, Cambridge, UK, pp 120–175

Wang H, Liao Q, Zhu X, Li J, Tian X (2010) Experimental studies of liquid droplet coalescence on the gradients. Surface J Superconduct Novel Magnet 23:1165–1168

Wang XD, Tian Y, Peng XF (2003) Self-aggregation of vapor–liquid phase transition. Progr Natl Sci (Chin) 13:281–286

Watson RGH, Birt DCP, Honour CW, Ash BW (1962) The promotion of dropwise condensation by montan wax I. Heat transfer measurements. J Appl Chem 12(12):539–546

Wenzel RN (1936) Resistance of solid surfaces to wetting by water. Ind Eng Chem 28:988–994

Wilmshurst R, Rose JW (1970) Dropwise condensation-further heat-transfer measurements. Proc 4th Int Heat Transf Conf 4:1–4

Wilmshurst R, Rose JW (1974) Dropwise and filmwise condensation of aniline ethandiol, and nitrobenzene. Proc 5th Int Heat Transf Conf 3:269–273

Wu M, Cubaud T, Ho CM (2004) Scaling law in a liquid drop coalescence driven by surface tension. Phys Fluids 16(7):51–54

Wu WH, Maa JR (1976) On the heat transfer in dropwise condensation. Chem Eng J 12:225–231

Wu Y, Yang C, Yuan X (2001) Drop distribution and numerical simulation of dropwise condensation heat transfer. Int J Heat Mass Transf 44:4455–4464

Yang CX, Wang LG, Yuan XG, Ma CF (1998) Dropwise condensation as typical fractal growth process. J Aerospace Power 13(3):272–276

Yoshida N, Abe Y, Shigeta H, Nakajima A, Ohsaki K, Watanable T (2006) Sliding behavior of droplet on flat polymer surface. J Am Chem Soc 128:743–747

Zhang DC, Lin ZQ, Lin JF (1986) New materials for dropwise condensation. Proc 8th Int Heat Transf Conf 4:1677–1682

Zhao H, Beysens D (1995) From droplet growth to film growth on a heterogeneous surface: condensation associated with a wettability gradient. Langmuir 11(2):627–634

Zhao Q, Burnside BM (1994) Dropwise condensation of steam on ion implanted condenser surfaces. Heat Recov Syst CHP 14:525–534

Zhao Q, Zhang DC, Lin JF (1991) Surface materials with dropwise condensation mode by ion implantation technology. Int J Heat Mass Transf 34:2833–2835

Zhao Q, Zhang DC, Lin JF, Wang GM (1996) Dropwise condensation on L-B film surface. Chem Eng Process 35:473–477

Zhao Q, Zhang DC, Zhu XB, Xu DQ, Lin ZQ, Lin JF (1990) Industrial application of dropwise condensation. Proc 9th Int Heat Transf Conf 4:391–394

Zhu X, Wang H, Liaon Q, Ding YD, Gu YB (2009) Experiments and analysis on self-motion behaviors of liquid droplets on gradient surfaces. Exp Therm Fluid Sci 33(6):947–954

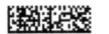